Quantum Space Theory With Application to Octonion Cosmology & Possibly To Fermionic Condensed Matter

Stephen Blaha Ph. D.
Blaha Research

Pingree-Hill Publishing

MMXXI

Rev. 00/00/01   August 21, 2021

To Margaret

Some Other Books by Stephen Blaha

All the Megaverse! Starships Exploring the Endless Universes of the Cosmos using the Baryonic Force (Blaha Research, Auburn, NH, 2014)

SuperCivilizations: Civilizations as Superorganisms (McMann-Fisher Publishing, Auburn, NH, 2010)

All the Universe! Faster Than Light Tachyon Quark Starships & Particle Accelerators with the LHC as a Prototype Starship Drive Scientific Edition (Pingree-Hill Publishing, Auburn, NH, 2011).

Unification of God Theory and Unified SuperStandard Model THIRD EDITION (Pingree Hill Publishing, Auburn, NH, 2018).

The Exact QED Calculation of the Fine Structure Constant Implies ALL 4D Universes have the Same Physics/Life Prospects (Pingree Hill Publishing, Auburn, NH, 2019).

Unified SuperStandard Theory and the SuperUniverse Model: The Foundation of Science (Pingree Hill Publishing, Auburn, NH, 2018).

Quaternion Unified SuperStandard Theory (The QUeST) and Megaverse Octonion SuperStandard Theory (MOST) (Pingree Hill Publishing, Auburn, NH, 2020).

Unified SuperStandard Theories for Quaternion Universes & The Octonion Megaverse (Pingree Hill Publishing, Auburn, NH, 2020).

The Essence of Eternity: Quaternion & Octonion SuperStandard Theories (Pingree Hill Publishing, Auburn, NH, 2020).

A Very Conscious Universe (Pingree Hill Publishing, Auburn, NH, 2020).

From Octonion Cosmology to the Unified SuperStandard Theory of Particles (Pingree Hill Publishing, Auburn, NH, 2020).

Beyond Octonion Cosmology (Pingree Hill Publishing, Auburn, NH, 2021).

Available on Amazon.com, bn.com Amazon.co.uk and other international web sites as well as at better bookstores (through Ingram Distributors).

CONTENTS

FIGURES and TABLES

Introduction

This book fleshes out the author's Quantum Space Theory: a Quantum Field Theory of Particles containing internal spaces. Fermion, scalar boson vector boson, and graviton space particles are formulated. Spinor indices and square spinor arrays play a crucial role in the theory. They lead to the dimension arrays of the ten spaces of Octonion Cosmology.

The unitary symmetry group on spinor indices has both global and local formulations. We show that they lead to a map from square spinor arrays to arrays of dimensions for octonion spaces.

The Octonion Spaces Spectrum is derived from a 20 space-time dimension point (thus giving upper limit to the octonion spaces spectrum). The spectrum is generated by repeated fermion-antifermion annihilation.

The Perturbation Theory of Quantum Space Theory is developed and applied to the creation of an octonion space from fermion-antifermion annihilation.

Analysis of spinor arrays leads to blocks of 16 fundamental fermions; and blocks of internal symmetries with irreducible representations totalling 16 dimensions. The blocking of 64 dimensions into fermion layers is also derived.

Quantum Space Theory leads (via Octonion Cosmology) to the author's NEWQUeST, NEWUTMOST, and NEWUST. It provides a complete derivation of Octonion Cosmology yielding the known features of elementary particles.

The author's Theory of Everything is thereby based on the Quantum Field Theory of Spaces. It has a clear, direct line of development based on the author's papers in the 1970's, and books in the Twenty-First Century.

The book suggests experimental tests of Octonion Cosmology. The book discusses the possible relation of the Octonion Spaces spectrum to the various SuperString theories.

Lastly, the book indicates the possible application of Quantum Space Theory to Fermionic Condensed Matter.

1. Quantum Field Theory of Fermion Space Particles

Quantum Field Theory[1] has had a long history of success in dealing with elementary particles. Its primary motivation is its ability to well describe the discrete nature of particles in a quantum framework as opposed to continuous matter.

Universes (and Megaverses, and so on) which have a multidimensional character, must be handled in a different manner—they have a particulate nature—but they also have additional structure requiring a more general formulation.[2] In this book we begin by creating a Quantum Space Theory for instances of particles with an inner space containing a space-time and a set of internal symmetries that evolves according to physical law.

The instances will be particles having child spaces with internal dimensions. The instances will have particle features and support interactions within a resident parent space based on a Lagrangian formulation. Thus the instances of the theory will exist within one space-time and yet the instances will have internal space-times of lower dimension that governs evolution within the instances' child spaces. See Fig. 1.1. Space instance dynamics are more complex because of the internal evolution within the child spaces.

This *modus operandi* supports the generation of spaces and instances within Octonion Cosmology as illustrated by Fig. 1.2.

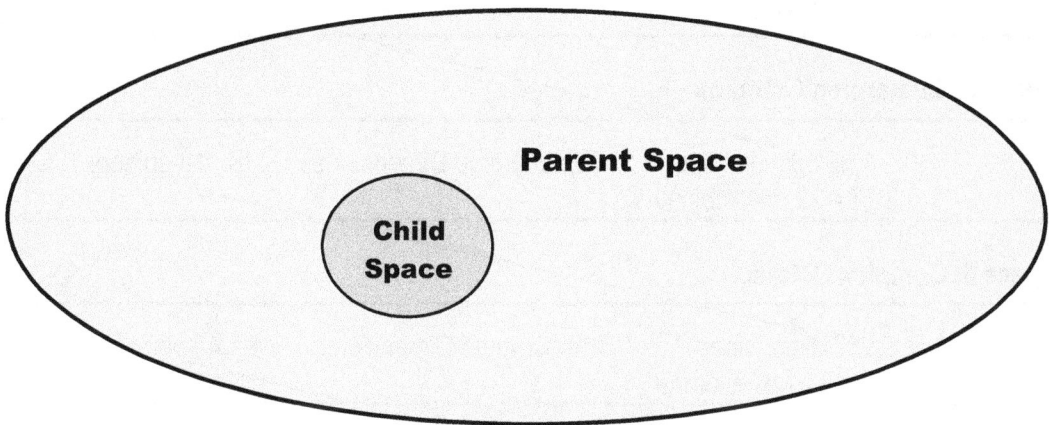

Figure 1.1. A space containing an instance that has an internal space of lower dimension. The space is itself within an instance of a higher space.

---

[1] Bjorken (1964) and (1965) are classic works in QFT. We generally follow their notation in this book and use the metric $g^{00} = +1$ with $g^{11} = g^{22} = g^{33} = -1$.
[2] We must distinguish between a space which is a multidimensional entity consisting of a space-time plus a set of internal symmetries, and an instance of the space that consists of mass and energy within the space-time governed by the internal symmetries and physical law. We shall use the word "space" and an instance interchangeably with the conext determining whether we are dealing with a space or an instance of the space. This procedure will avoid cumbersone verbiage while avoiding possible confusion.

God-Space – Space 0: Complex Octonion Octonion Octonion

1,048,576 dimensions	18 Space-time Dimensions	512 512-spinors
$2^{20/2} = 1024$ rows/columns		

Space 1: Octonion Octonion Octonion

262,144 dimensions	16 Space-time Dimensions	256 256-spinors
$2^{18/2} = 512$ rows/columns		

Space 2: Quarternion Octonion Octonion

65,536 dimensions	14 Space-time Dimensions	128 128-spinors
$2^{16/2} = 256$ rows/columns		

Space 3: Complex Octonion Octonion

16,384 dimensions	12 Space-time Dimensions	64 64-spinors
$2^{14/2} = 128$ rows/columns		

Space 4: Octonion Octonion

4,096 dimensions	10 Space-time Dimensions	32 32-spinors
$2^{12/2} = 64$ rows/columns		

Space 5: Quaternion Octonion

1,024 dimensions	8 Space-time Dimensions	16 16-spinors
$2^{10/2} = 32$ rows/columns		

Space 6: Complex Octonion

256 dimensions	4 Space-time Dimensions	4 4-spinors
$2^{8/2} = 16$ rows/columns		

Space 7: Octonion

64 dimensions	2 Space-time Dimensions	4 4-spinors
$2^{6/2} = 8$ rows/columns	(Built from four 16 dimension spaces of the 10 space spectrum)	

Figure 1.2. A sequence of fermion-antifermion annihilations generating instances in the set of Octonion Cosmology spaces.

We now turn to develop the Quantum Field Theory of Space Particles.

1.1 Free Space Particle Dirac Equation

In this section we will derive[3] the Space Fermion Particle Dirac equation using Lorentz boosts in r space-time dimensions of the wave function of a space particle at rest. We assume an r-dimension space-time with one time coordinate x^0 and $r - 1$ space components. In an r-dimension space time where r is an even integer, a spinor vector has $2^{r/2}$ coordinates, and a spinor array (which we introduce later) is a $2^{r/2} \times 2^{r/2}$ array.

We now turn to determine the form of the Dirac equation for an r-dimension space-time from Lorentz boosts.

1.1.1 Boosts in r Space-time Dimensions

The coordinates of inertial reference frames in r space-time dimensions are related by transformations of the r-dimensional inhomogeneous Lorentz group. To determine the form of its generators we consider infinitesimal "rotations" and translations. We denote translations by d^μ and homogeneous "rotations" by Λ^α_β. The identity rotation is $\Lambda_{16}{}^\alpha_\beta = \delta^\alpha_\beta$. An infinitesimal rotation can be expressed in the form

$$\Lambda^\alpha_\beta = \delta^\alpha_\beta + \omega^\alpha_\beta \tag{1.1}$$

The defining feature of Lorentz transformations is

$$\Lambda^T G \, \Lambda = G \tag{1.2}$$

where Λ is a "Lorentz" transformation expressed in matrix form, Λ^T is the transpose of Λ, and G is the metric $\eta_{\mu\nu}$ expressed as the $r \times r$ diagonal matrix: G = diag(1, -1, -1, ..., -1).. In the case of infinitesimal rotations eq. 1.2 becomes

$$(\delta^\alpha_\mu + \omega^\alpha_\mu)\eta_{\alpha\beta}(\delta^\beta_\nu + \omega^\beta_\nu) = \eta_{\mu\nu} \tag{1.3}$$

Using $\eta_{\alpha\beta}$ to lower indices and $\eta^{\alpha\beta}$ to raise indices we have

$$\omega_{\beta\mu} = \eta_{\alpha\beta}\omega^\alpha_\mu \tag{1.4}$$

so that eq. 1.2 implies the antisymmetry of $\omega_{\mu\beta}$

$$\omega_{\mu\beta} + \omega_{\beta\mu} = 0 \tag{1.5}$$

to leading order in ω.

The infinitesimal, unitary, linear Hilbert space transformation including a translation by d_μ, is

$$U(\omega, d) = I + \tfrac{1}{2} i \, \omega_{\alpha\beta} J^{\alpha\beta} - i d_\mu P^\mu \tag{1.6}$$

---

[3] These sections extend the discussion of universe particles of Blaha (2015a) and (2018e).

where the unitarity of U requires the $r(r - 1)/2$ operators $J^{\alpha\beta}$ and r operators P^{μ} to be hermitian, and eq. 1.5 requires $J^{\alpha\beta}$ to be antisymmetric in α and β. After some further conventional considerations, the following r-dimensional Poincaré group generator algebra commutation relations result with the metric $\eta_{16\mu\nu}$ and its inverse $\eta_{16}^{\mu\nu}$.

$$[J^{\alpha\beta}, J^{\kappa\lambda}] = i(\eta_{16}^{\alpha\kappa}J^{\beta\lambda} + \eta_{16}^{\lambda\alpha}J^{\kappa\beta} - \eta_{16}^{\beta\kappa}J^{\alpha\lambda} - \eta_{16}^{\lambda\beta}J^{\kappa\alpha}) \tag{1.7}$$

$$[P^{\alpha}, J^{\kappa\lambda}] = i(\eta_{16}^{\alpha\lambda}P^{\kappa} - \eta_{16}^{\alpha\kappa}P^{\lambda}) \tag{1.8}$$

$$[P^{\alpha}, P^{\kappa}] = 0 \tag{1.9}$$

The momentum operators are

$$\mathbf{P} = (P^0, P^1, \ldots, P^{r-1}) \tag{1.10}$$

The boost $(r - 1)$-vector is

$$\mathbf{K} = (J^{10}, J^{20}, \ldots, J^{(r-1)0}) \tag{1.11}$$

The angular momentum $(r - 1)$-vector is

$$\mathbf{J} = (J^{(r-2)(r-1)}, J^{(r-1)1}, J^{12}, J^{23}, J^{34}, \ldots, J^{(r-3)(r-2)}) \tag{1.12}$$

The form of an inhomogeneous r-dimensional Lorentz, Hilbert space transformation is

$$U(\mathbf{v}, \boldsymbol{\theta}, \mathbf{d}) = \exp[i\mathbf{a}\cdot\mathbf{K} + i\boldsymbol{\theta}\cdot\mathbf{J} - i\mathbf{d}\cdot\mathbf{P}] \tag{1.13}$$

where $\mathbf{a}$, $\boldsymbol{\theta}$ and $\mathbf{d}$ are complex.

A homogeneous Lorentz transformation $S(\mathbf{v}) = U(\mathbf{v}, \boldsymbol{\theta} = 0, \mathbf{d} = 0)$ on r-space-time dimension Dirac matrices (of size $2^{r/2} \times 2^{r/2}$ rows and columns) is[4]

$$S^{-1}(v)\gamma^{\nu}S(v) = \Lambda^{\nu}_{\mu}(v)\gamma^{\mu} \tag{1.14}$$

where $\Lambda^{\nu}_{\mu}(v)$ is the corresponding Lorentz coordinate transformation, and the inverse satisfies

$$S^{-1}(v) = \gamma^0 S^{\dagger}(v)\gamma^0 \tag{1.15}$$

where $\dagger$ signifies Hermitian conjugate.

---

[4] Blaha (2007b) gives the explicit form of s for four dimensions.

1.1.2 Generation of Fermion Space Particle Wave Function

We now derive the fermion space particle wave function using a Lorentz boost. We specify a generic positive energy plane wave solution of the Dirac equation for a space particle at rest with mass m as

$$\psi(x) = e^{-imt}W(0) \qquad (1.16)$$

with W(0) being a $2^{r/2} \times 2^{r/2}$ matrix (which is a generalization of the $2^{r/2}$ component spinor column vector for a normal fermion.)

$\psi(x)$ satisfies the momentum space Dirac equation for a space particle at rest:

$$(m\gamma^0 - m)e^{-imt}W(0) = 0 \qquad (1.17)$$

W(0) has the form $W(0) = \gamma^0$. If we apply S(v) we find

$$0 = S(v)(m\gamma^0 - m)e^{-imt}W(0) = [mS(v)\gamma^0 S^{-1}(v) - m]S(v)W(0)$$

We now require the Lorentz transformation to satisfy

$$mS(v)\gamma^0 S^{-1}(v) = g_{\mu\nu}p^\mu\gamma^\nu = \not{p} \qquad (1.18)$$

where $p^0 = (p^2 + m^2)^{\frac{1}{2}}$, $\mathbf{p} = \gamma m\mathbf{v}$, and $p = |\mathbf{p}|$ in r-dimension space-time. In addition we define

$$W(p) = S(v)W(0) \qquad (1.19)$$

Therefore the free space particle Dirac equation in momentum space has the form:

$$(\not{p} - m)e^{-ip\cdot x}W(p) = 0 \qquad (1.20)$$

NO m - p   m − m
where the exponential factor, "mt", is also boosted to p·x. Eq. 1.20 implies the free, coordinate space Dirac equation:

$$(i\gamma^\mu \partial/\partial x^\mu - m)\psi(x) = 0 \qquad (1.21)$$

1.1.3 Form of the Positive Frequency W Matrix

Eq. 1.18 determines the form of S(v) in terms of p:

$$S(v) = S(p) = (2m(m + p^0))^{-\frac{1}{2}}\gamma^0(\not{p} + m\gamma^0) \qquad (1.22)$$

and

$$S^{-1}(v) = S^{-1}(p) = (2m(m + p^0))^{-\frac{1}{2}}\gamma^0(\not{p} + m\gamma^0)\gamma^0 \qquad (1.23)$$

where p^0 is the energy. Note eqs. 1.22 and 1.23 implement eqs. 1.15 and 1.18.

1.1.3 Form of the Negative Frequency (Hole) W Matrix

The form of the Dirac equation for negative frequency is

$$(\not{p} + m)e^{-ip\cdot x}W'(p) = 0 \qquad (1.24)$$

Eq. 1.24 for the negative frequency Dirac equation requires a different Lorentz transformation $S'(p)$. The equation

$$mS'(p)\gamma^0 S'^{-1}(p) = -g_{\mu\nu}p^\mu\gamma^\nu = -\not{p} \qquad (1.25)$$

determines the form of $S'(v)$ in terms of p:

$$S'(p) = (2m(m - p^0))^{-\frac{1}{2}}\gamma^0(-\not{p} + m\gamma^0) \qquad (1.26)$$

and

$$S'^{-1}(p) = (2m(m - p^0))^{-\frac{1}{2}}\gamma^0(-\not{p} + m\gamma^0)\gamma^0 \qquad (1.27)$$

where p^0 is the energy. Consequently

$$W'(p) = S'(p)W(0) \qquad (1.28)$$

Therefore the negative frequency free space particle Dirac equation in momentum space has the form:

$$(\not{p} + m)e^{-ip\cdot x}W'(p) = 0 \qquad (1.29)$$

1.2 Free Lagrangian

We tentatively define the Lagrangian for the spin ½ space particle interacting with the fermion space particle to be:

$$\mathcal{L} = \overline{\psi}(i\partial_\mu\gamma^\mu - g'\Phi - m)\psi + \overline{\Phi}(i\partial_\mu\gamma^\mu - m')\Phi + \dots$$

Φ is a scalar space field with a pair of spinor indices. It is discussed in chapter 2. It is clear from the preceding section that ψ is a spinor field

1.3 Wave Particle Expansion

Having determined the form of $S(v)$ and $S'(p)$, and thus $W(p)$ and $W'(p)$ we can specify the form of an r space-time dimension fermion *space* wave function with one time dimension and r – 1 spatial dimensions. We will use the form of the Lorentz boost eq. 1.14 since it determines the spinor factors in the fermion wave function. We will also specify the form of a scalar boson (chapter 2) for a space particle using the form of the Lorentz boost. (Fermion-antifermion annihilation into the scalar boson will show that the scalar boson wave function "uses" the fermion spinor factors as an input.) The spinor arrays U and V have their size determined by r, which we call the *parent* space space-time dimension. The space-time dimension q is the *child* space space-time dimension.

The space fermion wave function for an r space-time dimension space fermion (with a q-dimension inner space-time) is[5]

$$\Psi_a(r, q, x) = \int d^{r-1}p(2\pi)^{-(r-1)}(m/p^0)^{\frac{1}{2}} \{exp^{-ip\cdot x} U_{a\beta}(r, q, p)b_\beta(p) +$$

$$+ exp^{ip\cdot x} V_{a\beta}(r, q, p)d^\dagger_\beta(p)\} \quad (1.30)$$

where a and β are spinor indices. Its Hermitian conjugate is

$$\Psi^\dagger_a(r, q, x) = \int d^{r-1}p(2\pi)^{-(r-1)}(m/p^0)^{\frac{1}{2}} \{exp^{ip\cdot x} b^\dagger_\beta(p)U^\dagger_{\beta a}(r, q, p) +$$

$$+ exp^{-ip\cdot x} d_\beta(p)V^\dagger_{\beta a}(r, q, p)\} \quad (1.31)$$

One may visualize the form of the spinor matrix U as a matrix with $2^{r/2}$ spinor columns, in which each column is a spinor with $2^{r/2}$ components. (Fig. 1.3) We view U as a matrix in spinor space. See chapter 5.

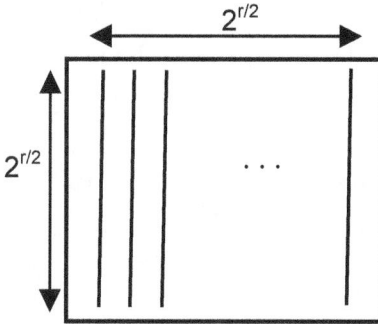

Figure 1.3. The U array with $2^{r/2}$ columns and $2^{r/2}$ rows. The V array has the same form.

Eq. 1.30 expresses Ψ as a sum over spins with a summation over β. For each value of β (column index) there is a creation/annihilation operator (b_β and $d_\beta^\dagger$) and a spinor vector $U_{\alpha\beta}(r, q, p)$ or $V_{\alpha\beta}(r, q, p)$ with indices labeled by α *as rows* within the spinor array. For each column spinor α labels the components within it.

We define spinor arrays ($2^{r/2} \times 2^{r/2}$) with

---

[5] This form makes β into the spinor index. It implements the close connection of the space feature and the spinors. Alternately an internal symmetry index could be introduced (as done in the scalar case in chapter 2) for the internal symmetry described in chapter 4. The sets of b_β and of d_β operators each constitute a fundamental q-number representation of $U(2^{r/2})$. See chapter 4.

$$U(r, q, p, m) = ((m + p^0)/2m)^{1/2}\gamma^0 S(p)$$
$$= (\not{p} + m)/2m \tag{1.32}$$

$$V(r, q, p, m) = ((m - p^0)/2m)^{1/2}\gamma^0 S'(p)$$
$$= (m - \not{p})/2m \tag{1.33}$$

Note that U and V are projections satisfying

$$U(r, q, p, m)^2 = U(r, q, p, m) \tag{1.34}$$
$$V(r, q, p, m)^2 = V(r, q, p, m)$$
$$U(r, q, p, m)V(r, q, p, m) = V(r, q, p, m)U(r, q, p, m) = 0 \tag{1.35}$$

Eq. 1.30 cleanly separates the positive and negative frequency parts of $\Psi(r, q, x)$. The spinor arrays are normalized column by column:

1.   The columns of U form an orthonormal set of $2^{r/2}$-vectors.
1.   The columns of V form an orthonormal set of $2^{r/2}$-vectors.

and

$$[\gamma^0 U(r, q, p, m)]^\dagger{}_{\alpha\beta} U(r, q, p, m)_{\beta\sigma} = [\gamma^0 U(r, q, p, m)]_{\alpha\sigma}$$
$$[\gamma^0 V(r, q, p, m)]^\dagger{}_{\alpha\beta} V(r, q, p, m)_{\beta\sigma} = [\gamma^0 V(r, q, p, m)]_{\alpha\sigma}$$
$$U(r, q, p, m)\gamma^0 U(r, q, p, m)^\dagger = U(r, q, p, m)\,\gamma^0 \tag{1.36}$$
$$V(r, q, p, m)\gamma^0 V(r, q, p, m)^\dagger = V(r, q, p, m)\,\gamma^0$$

$$U(r, q, p, m)_{\alpha\beta} U(r, q, p, m)^\dagger{}_{\beta\sigma} = (E/m)[U(r, q, p, m)\gamma^0]_{\alpha\sigma}$$
$$V(r, q, p, m)_{\alpha\beta} V(r, q, p, m)^\dagger{}_{\beta\sigma} = -(E/m)[V(r, q, p, m)\gamma^0]_{\alpha\sigma} \tag{1.37}$$

with $p^0 = E$. Note:

$$\gamma^0 U(r, q, p, m)^\dagger = U(r, q, p, m)\,\gamma^0 \tag{1.37a}$$

$$\gamma^0 V(r, q, p, m)^\dagger = V(r, q, p, m)\,\gamma^0 \tag{1.37b}$$

The creation and annihilation operators have a spinor index and satisfy fermion anticommutation relations:

$$\{b_\beta(p), b^\dagger{}_\sigma(p')\} = \delta_{\beta\sigma}\,\delta^{r-1}(p - p') \tag{1.38}$$
$$\{b_\beta(p), b_\sigma(p')\} = 0$$
$$\{b^\dagger{}_\beta(p), b^\dagger{}_\sigma(p')\} = 0$$

$$\{ d_\beta(p), d^\dagger{}_\sigma(p')\} = \delta_{\beta\sigma}\,\delta^{r-1}(p - p') \tag{1.39}$$

$$\{d_\beta(p), d_\sigma(p')\} = 0$$
$$\{d^\dagger_\beta(p), d^\dagger_\sigma(p')\} = 0$$

Note β and σ are spin labels for the $2^{r/2}$ spin states. The spin of the fermion is

$$s = 2^{r/2\,-\,2} - \tfrac{1}{2} \qquad (1.40)$$

In four space-time dimensions the *space* fermion has spin ½. The fermion spin in various spaces may be high. This is not a problem remembering that higher spin resonances appear on Regge trajectories for example. They are also not a problem in perturbation theory if one uses our Two Tier form of Quantum Field Theory that eliminates divergences in perturbation theory for any number of space-time dimensions. See Blaha (2005a).

1.4 Fermion Space Particle Interpretation

Fermion space particles have the space-time dependence specified by their Dirac equations. Their spinor behavior is embodied in the U(r, q, p) and U(r, q, p) arrays. In chapter 5 we will show the spinor arrays have a dual role: they specify the fermion spin state and they furnish the dimension arrays of the Octonion Spectrum of Spaces. In their second role they become a space, which itself has a space-time symmetry as well as internal symmetries.

1.5 Anticommutation Relations and Feynman Propagator

Fermion space particle wave functions have the anticommutation relation:

$$\{\Psi_a(r, q, t, \mathbf{x}), \Psi^\dagger_b(r, q, t, \mathbf{x}')\} = \iint d^{r-1}p\; d^{r-1}p'\delta^{r-1}(\mathbf{p} - \mathbf{p}')\{U_{a\alpha}(p)U^\dagger_{b\alpha}(p')\; e^{ip\cdot(x-x')} +$$

$$+ V_{a\alpha}(p)V^\dagger_{b\alpha}(p')\; e^{-ip\cdot(x-x')}\}(m^2/EE')^{\frac{1}{2}}\,(2\pi)^{-3}$$

$$= \gamma^0{}_{ab}\,\delta^{r-1}(\mathbf{x} - \mathbf{x}') \qquad (1.41)$$

with other anticommutators zero, and with $p^0 = E$ and using eqs. 1.37 with the r, q, and m arguments surpressed in U and V.

The Feynman propagator is:

$$(S_F(x' - x)\,\gamma^0)_{ab} = -i<0|\Psi_a(r, q, x')\gamma^0\Psi^\dagger_b(r, q, x)|0>\theta(t' - t) +$$
$$+ i<0|\,\Psi^\dagger_b(r, q, x)\,\gamma^0\Psi_a(r, q, x')|0>\theta(t - t')$$

$$= \iint d^{r-1}p\; d^{r-1}p'\delta^{r-1}(\mathbf{p} - \mathbf{p}')\{-i[U_{a\alpha}(p)\,\gamma^0 U^\dagger_{\alpha b}(p')e^{ip\cdot(x-x')}]\theta(t' - t) +$$

$$+ i\,[V_{a\alpha}(p)\,\gamma^0 V^\dagger_{\alpha b}(p')\,e^{-ip\cdot(x-x')}]\theta(t - t')\}(m^2/EE')^{\frac{1}{2}}(2\pi)^{-(r-1)}$$

$$= \int d^{r-1}p\,\{-ie^{ip\cdot(x-x')}\,[U(p)\gamma^0]_{ab}\,\theta(t' - t) +$$

$$+ ie^{-ip\cdot(x-x')}[V(p)\gamma^0]_{ab}\,\theta(t-t')\}(2\pi)^{-(r-1)}\,(m/E)$$

using eq. 1.36

$$= -i\int d^{r-1}p\,(m/E)\{e^{ip\cdot(x-x')}\,[U(p)\gamma^0]_{ab}\,\theta(t'-t)\,+$$

$$+e^{-ip\cdot(x-x')}[V(p)\,\gamma^0]_{ab}\,\theta(t-t')\}(2\pi)^{-(r-1)}$$

Or

$$S_F(x'-x)_{ab} = -i\int d^{(r-1)}p\{e^{-ip\cdot(x'-x)}U(p)_{ab}\,\theta(t'-t) + e^{ip\cdot(x'-x)}V(p)_{ab}\,\theta(t-t')\}(2\pi)^{-(r-1)}$$

(1.42)

$$= \int d^{(r-1)}p\,e^{-ip\cdot(x'-x)}\,(2\pi)^{-r}\,(\not p_{ab}+m)/(p^2-m^2+i\varepsilon)$$

Thus the Feynman propagator for space fermions is the same as that of "normal" fermions for r dimension space-time.

1.6 Possible Application to Fermionic Condensed Matter

Quantum Space Theory is clearly applicable to Octonion Cosmology. Due to the importance of spin dynamics in fermionic Condensed Matter it may also be applicable in that area of research. See Footnote 6 of chapter 2 for an indication of the importance of spin in the autrhor's previous work on vortices in Superfluid He^3. Chapter 5 brings out further indications of the implications of Quantum Space Theory in fermion dynamics.

2. Scalar Space Particle Quantum Field Theory

Second quantization of particle wave functions uses Bose-Einstein quantization for bosons and Fermi-Dirac quantization for Fermions. Their origin is based in observations of particle behavior and the Pauli Exclusion Principle.

In this chapter we will define a Lorentz spin 0 scalar space particle field with second rank spinor indices. We will find it convenient to use Bose-Einstein quantization in view of its scalar nature. But we will formulate it internally as having the spinor form of a space fermion[6] in order to reach the goal of having fermion-antifermion annihilation produce a scalar[7] space particle containing a space with an internal symmetry array.[8]

2.1 The Lagrangian and The Dirac Equation

We tentatively define the Lagrangian for the scalar space particle interacting with the fermion space particle to be:[9]

$$\mathcal{L} = \overline{\psi}(i\partial_\mu\gamma^\mu - g'\Phi - m)\psi + \overline{\Phi}(i\partial_\mu\gamma^\mu - m')\,\Phi\, + \ldots \qquad (2.1)$$

The free Lorentz scalar space particle Dirac equation is:

$$(i\partial_\mu\gamma^\mu - m')\,\Phi(r, q, x) = 0 \qquad (2.2)$$

It implies the usual scalar Klein-Gordon equation

$$(\partial_\mu\partial^\mu + m'^2)\,\Phi(r, q, x) = 0 \qquad (2.2a)$$

---

[6] There is a resemblance between Bose-Einstein even spin and Fermi-Dirac half integer spin phenomena. This is illustrated by the author's discovery (S. Blaha, Phys. Rev. Lett., **36**, 874 (1976)) of Fermi Superfluid He[3] quantized vortices, which resemble vortices in Bose superfluid helium.

[7] We take "scalar" to mean having Bose-Einstein statistics that allow any number of scalar particles with the same quantum numbers. Scalar particles do not obey the Pauli Exclusion Principle. Scalar refers to space-time not internal symmetry.

[8] Knowing that some will object to our quantization procedure the author acknowledges that he is *solely* responsible for his theoretical results. He has not received any "ideas", feedback, suggestions, or comments from others. Contrary to some rumors the author has had no academic affiliation in the past twenty years, and only an honorary affiliation in the preceding 17 years (1983-2000) with little contact with other physicists due to his commitments to other areas of research. In the past 10 years the author has not had any contact with other physicists. In the precding 10 years (2000-2010) there were a few, very brief, contacts of a merely social nature so that the author could pursue his original line of research unencumbered by preconceptions.

Blaha Research is an independent, privately-funded research institute founded in 2000 with the purpose of advancing human knowledge *expeditiously*.

[9] The Φ field we defined is a second rank spinor field and participates in the spinor dynamics of the model. If one chose to define a scalar field with no spinor indices then the spin dynasmics would be unaffected by it and thus not be relevant.

2.2 Scalar Space Particle Wave Function

Emulating the fermion wave function of the previous chapter, we choose the scalar space particle wave function with second rank spinor indices for an r space-time dimension space boson (with a q-dimension inner space-time) to be

$$\Phi_{ab}(r, q, x) = \int d^{r-1}p (2\pi)^{-(r-1)} (m'/p^0)^{1/2} \{\exp^{-ip\cdot x} U_{a\beta}(r, q, p, m')a_{\beta b}(p) +$$

$$+ \exp^{ip\cdot x} V_{a\beta}(r, q, p, m')c_{\beta b}^{\dagger}(p)\} \qquad (2.3)$$

where a, b, and β are spinor internal symmetry indices. Its Hermitian conjugate is

$$\Phi_{ab}^{\dagger}(r, q, x) = \int d^{r-1}p (2\pi)^{-(r-1)} (m'/p^0)^{1/2} \{\exp^{ip\cdot x} U_{a\beta}^{\dagger}(r, q, p, m')\, a_{\beta b}^{\dagger}(p) +$$

$$+ \exp^{-ip\cdot x} V_{a\beta}^{\dagger}(r, q, p, m')\, c_{\beta b}(p)\} \qquad (2.4)$$

The constant m' is the scalar particle mass as well as the energy of the created space, and U and V are the same as in fermion case. The spinor arrays, U and V, have $2^{r/2}$ columns and $2^{r/2}$ rows. The spinor arrays U and V have their size determined by r, which we call the *parent* space space-time dimension. The space-time dimension q is the *child* space space-time dimension. Note Φ is a second rank spinor. The sets of $a_{\beta b}$ and of $c_{\beta b}$ operators each constitute a q-number representation of $U(2^{r/2})$. See chapter 5.

The creation/annihilation operators satisfy the commutation relations:

$$[a_{jk}(p), a_{j'k'}^{\dagger}(p')] = \delta_{jj'}\delta_{kk'}\, \delta^{r-1}(p - p') \qquad (2.5)$$
$$[a_{jk}(p), a_{j'k'}(p')] = 0$$
$$[a_{jk}^{\dagger}(p), a_{j'k'}^{\dagger}(p')] = 0$$

plus similar commutation relations for $c_{jk}(p)$ and $c_{j'k'}^{\dagger}(p')$. The $a_{jk}(p)$ and $a_{j'k'}^{\dagger}(p')$ commute with $c_{jk}(p)$ and $c_{j'k'}^{\dagger}(p')$. There are two sets: 2^r operators $a_{jk}(p)$ and 2^r operators c_{jk}. Each set furnishes a representation of $U(2^{r/2})$.

2.3 Φ Commutation Relations

Scalar space particle wave functions have the commutation relations:

$$[\Phi_{ab}(r, q, t, \mathbf{x}), \Phi_{cd}^{\dagger}(r, q, t, \mathbf{x'})] = \iint d^{r-1}p\, d^{r-1}p'\delta^{r-1}(\mathbf{p} - \mathbf{p'})\{U_{a\beta}(p)U_{c\gamma}^{\dagger}(p')\delta_{\beta\gamma}\delta_{bd}e^{ip\cdot(x-x')} +$$
$$+ V_{a\beta}(p)V_{c\gamma}^{\dagger}(p')\, \delta_{\beta\gamma}\delta_{bd}\, e^{-ip\cdot(x-x')}\}(m^2/EE')^{1/2}\, (2\pi)^{-(r-1)}$$

$$= \gamma_{ac}^0\, \delta_{bd}\, \delta^{r-1}(\mathbf{x} - \mathbf{x'}) \qquad (2.6)$$

with $p^0 = E$ using eqs. 1.37 with the r, q, and m arguments of U and V not displayed.

2.4 Φ Feynman Propagator

The Feynman propagator is:

$$(S_F(x' - x) \gamma^0)_{acbd} = -i<0|\Phi_{ac}(r, q, x')\gamma^0\Phi^\dagger_{bd}(r, q, x)|0>\theta(t' - t) +$$
$$+ i <0| \Phi^\dagger_{bd}(r, q, x)\gamma^0\Phi_{ac}(r, q, x')|0>\theta(t - t')$$
$$= \iint d^{r-1}p \, d^{r-1}p'\delta^{r-1}(\mathbf{p} - \mathbf{p}')\{-i[U_{a\alpha}(p) \gamma^0 U^\dagger_{b\alpha}(p')\delta_{cd}e^{ip\cdot(x - x')}]\theta(t' - t) +$$

$$+ i \, [V_{a\alpha}(p) \gamma^0 V^\dagger_{b\alpha}(p')\delta_{cd}e^{-ip\cdot(x - x')}]\theta(t - t')\}(m'^2/EE')^{\frac{1}{2}}(2\pi)^{-(r-1)}$$

$$= \delta_{cd}\int d^{r-1}p \, \{-ie^{ip\cdot(x - x')} [U(p)\gamma^0]_{ab}\theta(t' - t) +$$

$$+ ie^{-ip\cdot(x - x')}[V(p)\gamma^0]_{ab} \theta(t - t')\}(2\pi)^{-(r-1)} (m'/E)$$

using eq. 1.36.

$$= -i\delta_{cd} \int d^{r-1}p \, (m'/E)\{e^{ip\cdot(x - x')} [U(p)\gamma^0]_{ab}\theta(t' - t) +$$

$$+e^{-ip\cdot(x - x')}[V(p) \gamma^0]_{ab}\theta(t - t')\}(2\pi)^{-(r-1)}$$

Or

$$S_F(x' - x)_{acbd} = -i \, \delta_{cd}\int d^{(r-1)}p\{e^{-ip\cdot(x' - x)}U(p)_{ab}\theta(t' - t) + e^{ip\cdot(x' - x)}V(p)_{ab} \theta(t - t')\}(2\pi)^{-(r-1)}$$

$$\qquad\qquad (2.7)$$

$$= \delta_{cd} \int d^r p \, e^{-ip\cdot(x' - x)} (2\pi)^{-r} (\not{p}_{ab} + m')/(p^2 - m'^2 +i\varepsilon)$$

Thus the form of the Feynman propagator for space scalars is almost the same as that for space fermions for r dimension space-time.

A scalar space particle has an r dimension space-time "skin" with r-momentum p, and contains a space which includes a q dimension internal space-time.

3. *Space* Vector Particle Quantum Field Theory

The quantum field theory of space spin 1 vector particles will be presented in this chapter in the *radiation* gauge:

$$A^0 = 0 \tag{3.1}$$
$$\nabla \cdot \mathbf{A} = 0$$

3.1 Lagrangian and Dynamic Equations

We define the Lagrangian for the spin 1 vector space particle interacting with the fermion space particle to be:

$$\mathcal{L} = \overline{\psi}[\gamma^\mu(i\partial_\mu - g'A_\mu \cdot T) - g\Phi - m]\psi + \overline{\Phi}(i\partial_\mu \gamma^\mu - m')\,\Phi\, - \tfrac{1}{4}\,F_{\mu\nu}F^{\mu\nu} + \ldots \tag{3.2}$$

where the T matrices are generators of the $U(2^{r/2})$ Dirac-Good group discussed in chapter 5, where the A_μ are the gauge vector fields of that group, and g' is its coupling constant. The $U(2^{r/2})$ group will be seen to be a local Yang-mills gauge group. The group acts on spinors. Therefore T has spinor indices. T also consists of 2^r generators: the r Dirac matrices and $2^r - 1$ other generators. T has the form T^i_{ab} for i = 1, …, 2^r and for a, b = 1, …, $2^{r/2}$.

The vector field in the radiation gauge satisfies

$$(\partial_\mu \partial^\mu + \xi^2)\, A^i(r, q, x) = 0 \tag{3.3}$$

for spatial coordinates i = 1, … , r – 1, where ξ is a small mass that we take to zero at the end of calculations (as done in QED infrared studies.) In order to introduce the U and V spinor arrays seen in previous chapters we require

$$(i\partial_\mu \gamma^\mu - \xi)\, A^i(r, q, x) = 0 \tag{3.4}$$

Note eq. 3.3 follows from eq. 3.4.

3.2 Vector Space Particle Wave Function

Emulating the scalar wave function of the previous chapter, we choose the vector space particle wave function for an r space-time dimension space gauge boson (with a q-dimension inner space-time) in the radiation gauge to be[10]

---

[10] With repeated indices summed.

$$\mathbf{A}^i_{ab}(r, q, x) = \int d^{r-1}p(2\pi)^{-(r-1)}(2p^0)^{-\frac{1}{2}}\Sigma_\lambda\varepsilon(p,\lambda)T^i_{a\alpha}\{exp^{-ip\cdot x}\, U_{\alpha\beta}(r, q, p, \xi)\, w_{\beta b}(p, \lambda) +$$

$$+\ exp^{ip\cdot x}\, V_{\alpha\beta}(r, q, p, \xi)\, z_{\beta b}^\dagger(p,\lambda)\} \tag{3.5}$$

where ε is the polarization unit vector; w and z are creation/annihilation operators with spinor indices—each taking $2^{r/2}$ values, λ takes $r - 2$ values, and U and V are similar to U and V in the fermion case. $\mathbf{A}^i_{ab}(r, q, x)$ is a vector of the spatial components of the vector field in the radiation gauge.

The spinor arrays, U and V, have $2^{r/2}$ columns and $2^{r/2}$ rows.. The spinor arrays U and V size is determined by r, the *parent* space space-time dimension. The space-time dimension q is the *child* space space-time dimension. Since U and V are not orthogonal for zero mass, we have introduced a "fictitious" mass ξ which is taken to zero at the end of calculations.

The Hermitian conjugate is

$$\mathbf{A}^i_{ab}(r, q, x)^\dagger = \int d^{r-1}p(2\pi)^{-(r-1)}(2p^0)^{-\frac{1}{2}}\Sigma_\lambda\varepsilon(p,\lambda)T^i_{a\alpha}\{exp^{ip\cdot x}\, U_{\alpha\beta}(r, q, p, \xi)^\dagger\, w_{\beta b}^\dagger(p, \lambda) +$$

$$+\ exp^{-ip\cdot x}\, V_{\alpha\beta}(r, q, p, \xi)^\dagger z_{\beta b}(p,\lambda)\} \tag{3.6}$$

The creation/annihilation operators satisfy the commutation relations:

$$[w_{jk}(p, \lambda), w_{j'k'}^\dagger(p', \lambda')] = \delta_{jj'}\delta_{kk'}\,\delta_{\lambda\lambda'}\,\delta^{r-1}(p - p') \tag{3.7}$$
$$[w_{jk}(p, \lambda), w_{j'k'}(p', \lambda')] = 0$$
$$[w_{jk}^\dagger(p, \lambda), w_{j'k'}^\dagger(p', \lambda')] = 0$$
$$[z_{jk}(p, \lambda), z_{j'k'}^\dagger(p', \lambda')] = \delta_{jj'}\delta_{kk'}\delta_{\lambda\lambda'}\delta^{r-1}(p - p')$$
$$[z_{jk}(p, \lambda), z_{j'k'}(p', \lambda')] = 0$$
$$[z_{jk}^\dagger(p, \lambda), z_{j'k'}^\dagger(p', \lambda')] = 0$$

with zero commutators between w and z operators.

For fixed λ, there are 2^r operators $w_{jk}(p, \lambda)$ and 2^r operators.Each set of $w_{jk}(p, \lambda)$ and of $z_{jk}(p, \lambda)$ is a q-number representation of $U(2^{r/2})$.

3.3 A Feynman Propagator

The Feynman propagator for the μth and ν^{th} space-time components of $\mathbf{A}^i_{ab}(r, q, x)$ is

$$(S_F(x' - x)\, \gamma^0)^{ijuv}_{acbd} = -i<0|\mathbf{A}^{iu}_{ac}(r, q, x')\gamma^0\mathbf{A}^{jv\dagger}_{bd}(r, q, x)|0>\theta(t' - t) +$$
$$+\ i<0|\ \mathbf{A}^{jv\dagger}_{bd}(r, q, x)\gamma^0\mathbf{A}^{iv}_{ac}(r, q, x')|0>\theta(t - t')$$

$$= \iint d^{r-1}p\ d^{r-1}p'\delta^{r-1}(\mathbf{p} - \mathbf{p}')\ \Sigma_\lambda\varepsilon^u(p,\lambda)\varepsilon^v(p',\lambda)T^i_{a\alpha}T^i_{b\kappa}\{-i[U_{\alpha\beta}(p)\ \gamma^0 U^\dagger_{\kappa\beta}(p')\delta_{cd}e^{ip\cdot(x - x')}]\theta(t' - t) +$$

$$+ i [V_{\alpha\beta}(p) \gamma^0 V^\dagger_{\kappa\beta}(p')\delta_{cd}e^{-ip\cdot(x-x')}]\theta(t-t')\}(\xi'^2/EE')^{\frac{1}{2}}(2\pi)^{-(r-1)}$$

$$= T^i_{a\alpha}T^j_{b\kappa} \delta_{cd}\int d^{r-1}p \, \Sigma_\lambda\varepsilon^u(p,\lambda)\varepsilon^v(p',\lambda)\{-ie^{ip\cdot(x-x')} [U(p)\gamma^0]_{\alpha\kappa}\theta(t'-t) +$$

$$+ ie^{-ip\cdot(x-x')}[V(p)\gamma^0]_{\alpha\kappa} \theta(t-t')\}(2\pi)^{-(r-1)} (\xi/E)$$

$$= -i\delta_{cd} \, T^i_{a\alpha}T^j_{b\kappa} \int d^{r-1}p \, (\xi/E) \Sigma_\lambda\varepsilon^u(p,\lambda)\varepsilon^v(p,\lambda)\{e^{ip\cdot(x-x')} [U(p)\gamma^0]_{\alpha\kappa}\theta(t'-t) +$$

$$+ e^{-ip\cdot(x-x')}[V(p) \gamma^0]_{\alpha\kappa}\theta(t-t')\}(2\pi)^{-(r-1)}$$

$$= \delta_{cd} \, T^i_{a\alpha}T^j_{b\kappa} \, D(d/dx) \, S_F(x' - x) \gamma^0)_{\alpha\kappa} \tag{3.8}$$

with

$$D^{uv} = \Sigma_\lambda\varepsilon^u(d/dx, \lambda)\varepsilon^v(d/dx, \lambda)$$

and

$$S_F(x' - x) \gamma^0)_{\alpha\kappa} = -i \int d^{r-1}p \, (\xi/E) \{e^{ip\cdot(x-x')} [U(p)\gamma^0]_{\alpha\kappa}\theta(t'-t) +$$
$$+ e^{-ip\cdot(x-x')}[V(p) \gamma^0]_{\alpha\kappa}\theta(t-t')\}(2\pi)^{-(r-1)}$$

$$= \int d^r p \, e^{-ip\cdot(x'-x)} (2\pi)^{-r} (\not{p}_{\alpha\kappa} + \xi)/(p^2 - \xi^2 +i\varepsilon)$$

Thus the form of the Feynman propagator for space vectors is almost the same as that for space fermions for r dimension space-time.

4. Space Graviton Quantum Field Theory

Our consideration of the fermion, scalar boson, and vector boson space fields leads to the possibility of graviton space fields. In a weak field approximation we define a metric tensor with two spinor indices a and b:

$$g_{ab}{}^{\mu\nu}(r, q, x) \cong \eta_{ab}{}^{\mu\nu}(r, q, x) + h_{ab}{}^{\mu\nu}(r, q, x) \qquad (4.1)$$

Following the conventional approach we arrive at the dynamic equation:

$$-\tfrac{1}{2} \,\square\, (h_{ab}{}^{\mu\nu} - \tfrac{1}{2} h\, \eta_{ab}{}^{\mu\nu}) = -16\pi G T_{ab}{}^{\mu\nu} \qquad (4.2)$$

based on the

$$R_{\mu\nu ab} - \tfrac{1}{2}\, R_{\alpha\beta} g_{\mu\nu\beta b} = 8\pi G T_{\mu\nu ab}$$

The spin 2 $h_{ab}{}^{\mu\nu}(r, q, x)$ field can be expanded to

$$h^{\mu\nu}{}_{ab}(r, q, x) = \int d^{r-1}p (2\pi)^{-(r-1)} (2p^0)^{-\tfrac{1}{2}} \Sigma_\lambda \varepsilon^{\mu\nu}(p,\lambda) \{ \exp^{-ip\cdot x} U_{\alpha\beta}(r, q, p, \xi)\, w_{\beta b}(p, \lambda) +$$

$$+\ \exp^{ip\cdot x} V_{\alpha\beta}(r, q, p, \xi)\, z_{\beta b}{}^{\dagger}(p,\lambda) \} \qquad (4.3)$$

with quantities defined similarly to the previous.

4.1 Space Graviton Propagator

The free Feynman propagator for space gravitons is:

$$(S_F(x' - x)\, \gamma^0)^{ijuv\rho\sigma}{}_{acbd} = -i<0|h^{uv}{}_{ac}(r, q, x')\gamma^0 h^{\rho\sigma\dagger}{}_{bd}(r, q, x)|0>\theta(t' - t) +$$

$$+\ i <0|\ h^{\rho\sigma\dagger}{}_{bd}(r, q, x)\gamma^0 h^{uv}{}_{ac}(r, q, x')|0>\theta(t - t')$$

$$= \delta_{cd}\, D^{uv\rho\sigma}(d/dx)\, S_F(x' - x)\, \gamma^0)_{ab} \qquad (4.4)$$

using

$$D^{uv\rho\sigma} = \Sigma_\lambda \varepsilon^{uv}(d/dx, \lambda)\varepsilon^{\rho\sigma}(d/dx, \lambda)$$

and

$$S_F(x' - x)\, \gamma^0)_{ab} = -i \int d^{r-1}p\, (\xi/E)\, \{e^{ip\cdot(x - x')}\, [U(p)\gamma^0]_{ab}\theta(t' - t) +$$

$$+ e^{-ip\cdot(x - x')}[V(p)\, \gamma^0]_{ab}\theta(t - t')\}(2\pi)^{-(r-1)}$$

$$= \int d^r p\, e^{-ip\cdot(x' - x)}\, (2\pi)^{-r} (\rlap{/}{p}_{ab} + \xi)/(p^2 - \xi^2 +i +i\varepsilon)$$

where ξ is an infinitesimal mass that goes to zero as in previous cases.

4.2 Spin Connection and Palatini-like Formulation of General Relativity

The space graviton formulation invites comparison to spin connections and the Palatini formulation of General Relativity. The Palatini formalism begins with transforming γ-matrices with a *vierbein*:

$$\gamma^{\mu}(x) = \gamma^{k} e^{\mu}_{k}(x) \qquad (4.5)$$

Thereby the connection to the present work is lost. Chapter 5 points out the importance of the Dirac-Good group for establishing the transformation of spinor arrays to dimension arrays.

However, the use of a *vierbein* development brings out similarities and suggests the use of a combined Palatini and Dirac-Good group (chapter 5) approach might be of some interest. We will pursue this possibility at another time.

4.3 Fermion Space Formulation vs. Boson Space Formulation

Comparing the Space Field Theory formulations for fermions with the formulation for bosons (scalar, vector and spin 2) we note the fermion formulation uses vector creation/annihilation operators (b_{β} and $d^{\dagger}_{\beta}$) with one index, while the boson formulations use second rank creation/annihilation operators ($c_{\beta b}$, $c_{\beta b}^{\dagger}$, $w_{\beta b}$, $z_{\beta b}^{\dagger}$) with two indices.[11]

---

[11] Creation/annihilation operators with no index are ruled out since field commutations would be unacceptable.

5. U($2^{r/2}$) Symmetry Groups

The space-time of an octonion space may contain a subspace (and perhaps more than one subspace). These spaces originate in the spinor spaces of fermions, scalar bosons, and vector bosons.

The space particle wave functions have two symmetries: one symmetry based on momentum conservation, and a spinor space symmetry based on symmetry groups of Dirac matrices. The second symmetry is the more important for present purposes. This symmetry depends directly on the space-time dimension of the space-time.

The ten spaces of Octonion Cosmology *each* have an associated U($2^{r/2}$) symmetry group for space-time dimension r that transforms the Dirac matrices of their space-time. In this chapter we explore the symmetry structure of each space of Octonion Cosmology.

5.1 Symmetry Group Related to Momentum Conservation

Examining the calculation of a scalar space particle from fermion-antifermion annihilation, eq. 3.3, we see the fermion and antifermion momenta combine to give the momentum of the scalar space particle:

$$V \sim (\not{p}_+) \text{ "$\times$" } U \sim (\not{p}_-) \rightarrow U \sim (\not{p}_+ + \not{p}_-) \tag{5.1}$$

because of momentum conservation at the interaction vertex. Conservation may be viewed at the level of Lorentz group boosts:

$$S(p_+) \text{ "$\times$" } S(p_-) \rightarrow S(p_+ + p) \tag{5.2}$$

using eq. 1.22 for Lorentz group boosts.

Momentum conservation then leads to an additive abelian symmetry of the spinor arrays. If we define the traceless spinor arrays:

$$U_t(p) = \not{p} \tag{5.3}$$
$$V_t(p) = \not{p}$$

then

$$U_t(p) \text{ "$\times$" } U_t(s) = U_t(p + s) \tag{5.4}$$
$$V_t(p) \text{ "$\times$" } V_t(s) = V_t(p + s)$$

5.2 Symmetry Group of Dirac Matrices

The second symmetry group that we will consider is a Dirac γ matrix symmetry in a space with an *even* space-time dimension r. The group is a U($2^{r/2}$) unitary symmetry group. We will call it the r-dimension Dirac-Good group.[12] It deals with the r γ-matrices of this space, and its 2^r matrices: each with $2^{r/2}$ rows and $2^{r/2}$ columns.

Dirac-Good group transformations satisfy:

$$\gamma'^{\mu} = U\gamma^{\mu}U^{-1} \qquad (5.5)$$

Hermiticity requires

$$\gamma'^{0\dagger} = \gamma'^0$$
$$\gamma'^{i\dagger} = -\gamma'^i \qquad (5.6)$$

where

$$U^{-1} = U^{\dagger} \qquad (5.7)$$

The γ'^{μ} matrices are physically equivalent to the Dirac matrices of space-time dimension r. Their anticommutation relations are the same as those of the γ^{μ} Dirac matrices.

5.3 Gauge Theory with Local γ-Matrix Transformations

The γ-matrices of each space initially start as constant. The Dirac-Good group performs global (coordinate independent) transformations on the γ-matrices as shown above.

If we define a Yang-Mills extension of the Dirac-Good group to enable local transformations, then we require a gauge field and interaction of the form:

$$I_{ab} = g' \overline{\psi}_a\gamma^{\mu}{}_{ab}A_{\mu bc}\cdot T_{cd}\psi_d \qquad (3.2)$$

Then local gauge transformations of Dirac matrices exist of the form

$$\gamma'^{\mu}(x) = U\gamma^{\mu}(x)U^{-1} \qquad (5.8)$$

Thus eq. 3.2 is a Yang-Mills lagrangian. It also has the term:

$$-\tfrac{1}{4} F_{\mu\nu}F^{\mu\nu} \qquad (5.9)$$

5.4 Map to Dimension Arrays

The spinor arrays, U and V, that appear in the previous chapters have been shown to have the form of Cayley arrays, and to lead to the dimension arrays of NEWQUeST (NEWUST) and NEWUTMOST for our universe and the Megaverse.[13]

The single index, fermion, creation/annihilation operators (b_{β} and $d^{\dagger}_{\beta}$) form a fundamental representation of the Dirac-Good group $U(2^{r/2})$. We noted some time ago that the form of the spinor arrays closely resembles the form of the NEWQUeST and QUeST internal symmetry group pattern. This resemblance gave us a basis for the

---

[12] The case of 4-dimension space-time was presented in R. H. Good, Jr., Rev. Mod. Phys **27**, 187 (1955). The generalization to r –dimension space-time is direct. We assume that the space-time dimension is an even number.
[13] Blaha (2021c) and (2021e) and earlier books.

connection between spinor arrays and dimension arrays. (Chapters 9 and 11 describe the resemblance in detail.) We now can assert the principle:

Dimension arrays have the same form as spinor arrays in each of the 10 spaces of Octonion Cosmology. The proof is in eq. 7.1 below. It shows the generation of the sequence of spaces from fermion-antifermion annihilation.

Fermion creation/annihilation operator indices become dimension array indices.

5.5 Importance of Spin ½

The central role of fermion-antifermion annihilation in generating spaces raises the question of a yet deeper basis of Octonion Cosmology in something like Twistor Theory or SuperString Theory.

6. Space Particle Perturbation Theory

The general form of perturbation theory is the same except for self-evident changes in such quantities as the number of dimensions. Bjorken (1965) has the traditional approach. However higher dimension calculations have divergences that cannot be eliminated by standard 4-dimension renormalization. The author's Two Tier approach,[14] which defines quantum coordinates, is based on

$$X^\mu(y) = y^\mu + i\, Y^\mu(y)/M_c^{\,2} \qquad (6.1)$$

where $Y^\mu(y)$ is a QED-like vector boson field, M_c is an extremely large mass, and the y^μ are the underlying coordinates. The two tiers of coordinates are the quantum coordinates $X^\mu(y)$ and the c-number coordinates y^μ. Two Tier coordinates eliminate *all* divergences (including fermion triangle divergences) in *all* dimensions.

Thus a *completely finite* perturbation theory is supported for all space-time dimensions r. See Blaha (2005a) for the Two Tier form of perturbation theory. It is similar to standard perturbation theory in most respects. *Its "low energy" limit for calculations yields conventional quantum field theory results.*

6.1 Fermion-Antifermion Annihilation into Scalar Space Particle

In this section we consider the creation of a scalar space particle[15] by off shell[16] fermion-antifermion annihilation. We will use the model Lagrangian density:[17]

$$\mathcal{L} = \overline{\psi}(i\partial_\mu \gamma^\mu - g\Phi - m)\psi + \overline{\Phi}(i\partial_\mu \gamma^\mu - m')\,\Phi \;\ldots \qquad (6.2)$$

where Φ is a space boson field array whose internal symmetry dimensions have the same number of rows and columns as the Dirac γ matrices. There is an interaction between the fermion spin and the internal symmetries of Φ that is analogous to the spin-orbit interaction in atomic physics.

The resulting lowest order annihilation process appears in Fig. 6.1. The S-matrix element is

$$S_{fi}^{\;pair} = N\, \delta^r(p_+ + p_- - p_s)\, V^\dagger(r, q, p_+, m)\gamma^0\, U(r, q, p_+ + p_-, m')U(r, q, p_-, m)$$
$$(6.3)$$

---

[14] See Blaha (2005a)

[15] We may perform the same calculation with a vector space particle (chapter 3) and obtain similar results.

[16] Off shell is required for energy-momentum conservation. An onshell process would require two scalar spaces to be created.

[17] The generation of dimension arrays could also be based on fermion-antifermion annihilation into a gauge Dirac-Good vector boson. (of chapter 3)

where N is a constant. S_{fi}^{pair} is an array of spin components. Individual S-matrix elements may be obtained by selecting specific in and out spins.

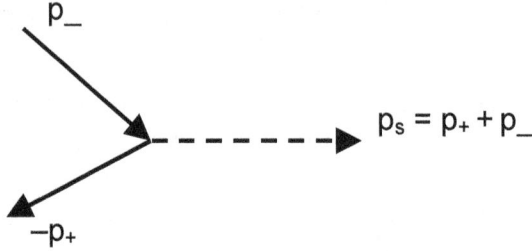

Figure 6.1. Diagram for the offshell annihilation of a space fermion – space antifermion pair to produce a space scalar particle.

Upon substituting for U and V we obtain;

$$S_{fi}^{\ pair} = N\ \delta^r(p_+ + p_- - p_s)\ (m - p\!\!\!/_+)^\dagger \gamma^0\ (p\!\!\!/_+ + p\!\!\!/_- + m')(p\!\!\!/_- + m)/(8m^2 m')$$

$$= N\ \delta^r(p_+ + p_- - p_s)\ \gamma^0(m - p\!\!\!/_+)(p\!\!\!/_- + m)/(8mm') \tag{6.4}$$

$$= N\ \delta^r(p_+ + p_- - p_s)\ \gamma^0 V(r, q, p\!\!\!/_+)\ U(r, q, p\!\!\!/_-)/2$$

$$= N\ \delta^r(p_+ + p_- - p_s)\ V^\dagger(r, q, p\!\!\!/_+)\gamma^0 U(r, q, p\!\!\!/_-)/2$$

$$= N\ \delta^r(p_+ + p_- - p_s)\ \overline{V}(r, q, p\!\!\!/_+)U(r, q, p\!\!\!/_-)/2 \tag{6.5}$$

$$= N\ \delta^r(p_+ + p_- - p_s)\ ((m - p_+^{\ 0})(m + p_-^{\ 0}))^{\frac{1}{2}}/(2m)\ S'(p_+)\gamma^0 S(p_-) \tag{6.6}$$

Comments:

1. $S_{fi}^{\ pair}$ is independent of the Φ mass m'.

2. If $p_- = p_s$ then $S_{fi}^{\ pair} = 0$ by eq. 1.35.

3. Cayley-Dickson numbers beyond octonions have non-trivial zero divisors where $n_1 \cdot n_2 = 0$, and n_1 and n_2 are non-zero Cayley-Dickson numbers. For each type of Cayley-Dickson number there are a *finite* number of non-trivial zero divisors.

    Arrays of Cayley-Dickson numbers were considered by the author in Blaha (2021a) – (2021c). The arrays (m - p\!\!\!/) and (p\!\!\!/ + m) are each an array of

Cayley numbers for each value of p. Together they form an *infinite* set of non-trivial zero divisors.

4. The space scalar particle internal symmetry array is generated by the fermion-antifermion annihilation. The form of the space scalar particle internal symmetry array $(\not{p}_+ + \not{p}_- + m)/(2m)$ follows from momentum conservation and leads to the internal symmetries of the octonion space spectrum.

7. Specification of Octonion Spaces

The space particles that we defined in earlier chapters support the Spectrum of Octonion Spaces that we developed in earlier books. We viewed the spectrum as originating in a cascade of fermion-antifermion annihilations as pictured in Fig. 7.2. We can express the cascade of spaces in terms of the scalar space particle wave function as

$$
\begin{array}{ll}
\mathbf{r} \quad \mathbf{q} & \\
\Phi(20,\ 18,\ m_{20},\ x) & (7.1)\\
\Phi(18,\ 16,\ m_{18},\ x) & \\
\Phi(16,\ 14,\ m_{16},\ x) & \\
\Phi(14,\ 12,\ m_{14},\ x) & \\
\Phi(12,\ 10,\ m_{12},\ x) & \\
\Phi(10,\ 8,\ m_{10},\ x) & \\
\Phi(8,\ 6,\ m_{8},\ x) & \\
\Phi(6,\ 4,\ m_{6},\ x) & \\
\Phi(4,\ 2,\ m_{4},\ x) & \\
\Phi(2,\ 0,\ m_{2},\ x) & \\
\end{array}
$$

The spaces, with $q = r - 2$, have space-time dimensions arranged to yield the Cayley-Dickson number pattern in Fig. 7.1. The sequence of space arrays is listed in Fig. 7.1. The masses $m_{20} \gg m_{18} \gg m_{16} \gg \ldots \gg m_2$ are the energies of the created spaces.

Cayley-Dickson Number n	Parent Space-time Dimension r	Total Dimension Array d_C	Child Space-time Dimensions q
10	20	1024×1024	18
9	18	512×512	16
8	16	256×256	14
7	14	128×128	12
6	12	64×64	10
5	10	32×32	8
4	8	16×16	6
3	6	8×8	4
2	4	4×4	2
1	2	2×2	0

Figure 7.1. Cayley-Dickson number n, and the corresponding tnumbers of total dimensions and of space-time dimensions. The number n = 1 corresponds to complex numbers. The number n = 10 corresponds to complex octonion octonion octonion numbers (with 1024 components).

The regularity of the list reflects the choice of space-time dimensions in eq. 7.1. The complete set of ten spaces is generated by nine nested fermion-antifermion annihilations. Since time in our universe does not exist outside of the universe, there is no issue of the amount of time required.

The cascade can generate side branches as shown in Fig. 7.2 below from Blaha (2021b). Sibling spaces are allowed—especially since we specify Bose-Einstein quantization of Φ.

Our Quantum Field Theory of Spaces supports, and is consistent with, the Octonion Cosmology Spectrum of Spaces.

A COSMOS

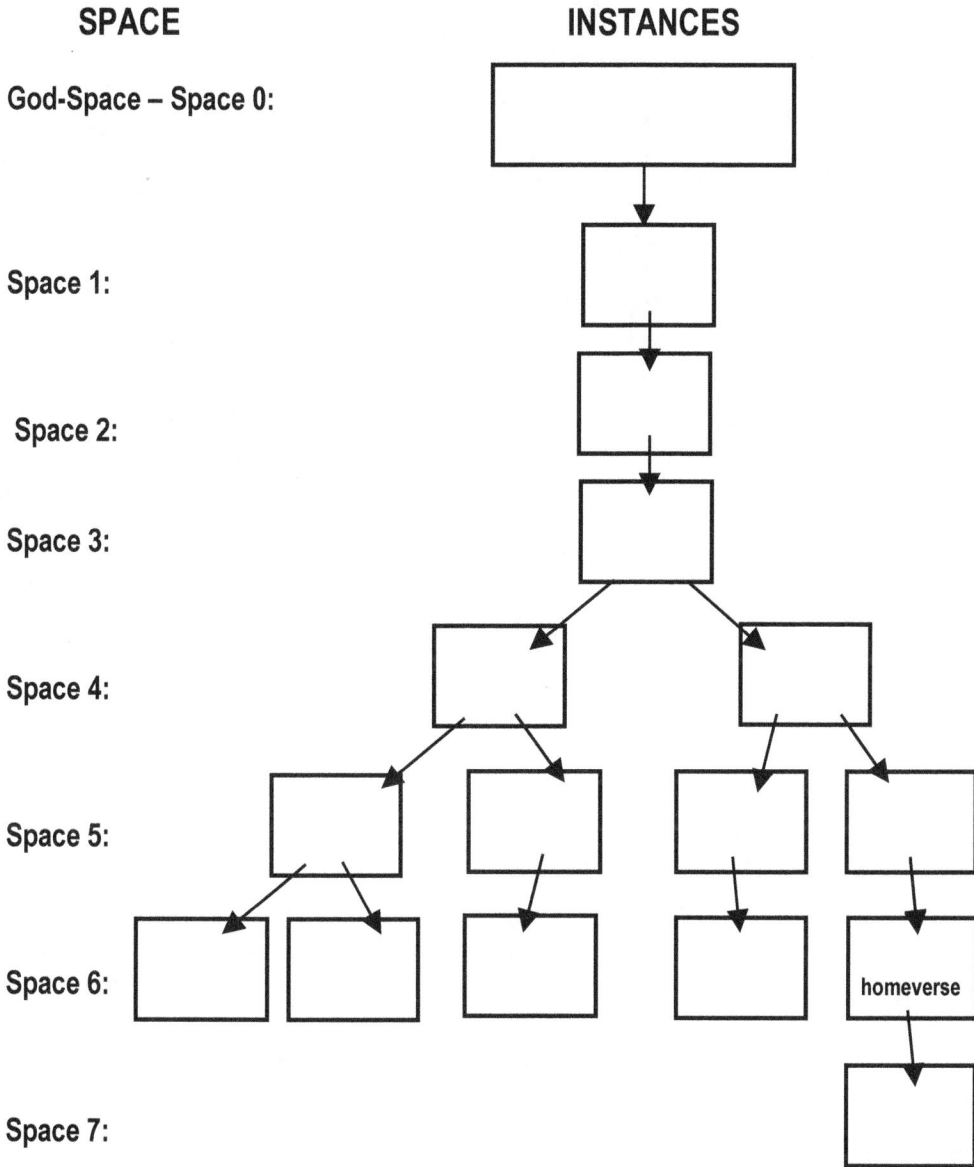

Figure 7.2. A hierarchy of instances leading from the God-Space (Cayley space 10) to "homeverse" – our designation for our universe. The homeverse is shown to contain one space 7 instance for illustration purposes. Space 7 is viewed here as combining spaces 7, 8, 9 and 10, which are part of the 10 space form of the spectrum. The homeverse has one "sibling" and three "cousin" universes. *The entire hierarchy resides in God-Space because the inheritance stems from the God-Space instance as parts of it. Other universes can be "reached" through the God-Space instance if a mode of transportation existed.*

8. Octonion Spaces Spectrum

The octonion spaces spectrum of Fig. 7.1 specifies ten spaces. Fig. 8.1 specifies it in more detail.

THE TEN OCTONION SPACES SPECTRUM

Octonion Space Number O_s	Cayley-Dickson Number n	Cayley Number Dimension d_c	Dimension Array Total d_d	Space-time-Dimension q	Fermion Spinor Array Total d_s	Cayley Number "Name"
0	10	1024	1024×1024	18	512×512	Complex Octonion Octonion Octonion
1	9	512	512×512	16	256×256	Octonion Octonion Octonion
2	8	256	256×256	14	128×128	Quaternion Octonion Octonion
3	7	128	128×128	12	64×64	Complex Octonion Octonion
4	6	64	64×64	10	32×32	Octonion Octonion
5	5	32	32×32	8	16×16	Quaternion Octonion
6	**4**	**16**	**16×16**	**6^{18}**	**8×8**	**Complex Octonion**
7	3	8	8×8	4	4×4	Octonion
8	2	4	4×4	2	2×2	Quaternion
9	1	2	2×2	0	1×1	Complex

Figure 8.1. The Octonion Cosmology ten space spectrum. The space for our universe, number 6, (Cayley number 4) is in bold type.

The various columns in Fig. 8.1 are related through the equations:

$$q = 2n - 2 \tag{8.1}$$

for the number of space-time dimensions and the Cayley number n. The Cayley number dimension d_c and the corresponding total size of the dimension array d_d are related by

$$d_c = 2^n \tag{8.2}$$

$$d_d = 2^{2n} \tag{8.3}$$

In addition

$$d_d^{(n)} = d_s^{(n+1)} \tag{8.4}$$

---

[18] The space-time dimensions become 4 as in the Unified SuperStandard Theory (UST) through either transfer of dimensions to internal symmetries or by the compactification of two dimensions. As a result the pattern of fermion-antifermion annihilations producing spaces 7, 8, and 9 is disrupted.

where n is the Cayley-Dickson number of a space and n + 1 is the Cayley number of the space above it. The quantity $d_s^{(n+1)}$ is the total size of the $(n + 1)^{th}$ fermion spinor array.

We also see the total number of spinor array components d_s for a space is

$$d_s = 2^{2n-2} \tag{8.5}$$
$$d_s = 2^r \tag{8.6}$$

implying

$$q = 2n - 2 \tag{8.7}$$

The spinor array's rows and columns sizes are

$$\text{Number of components in a spinor column} = 2^{n-1} \tag{8.8}$$

Consequently the fermion spin s is

$$s = (2^{n-2} - 1)/2 \tag{8.9}$$

These relations set the internal symmetry and space-time dimensions of the n > 4 spaces to those in Fig. 8.1.

8.1 Effect of Two Compacted Space 6 Dimensions

The compactification of 6 dimension space-time to 4 dimensions (space number 6) distorts the spectrum in Fig. 8.1 since the spaces 7, 8, and 9 generated by fermion-antifermion annihilation are different. See Fig. 8.2 for a modified Octonion Spaces Spectrum.

Modified *TEN* OCTONION SPACES SPECTRUM

Spectrum Number

	Coordinate Cayley Type	Dimension of a Coordinates	Dimension Array Size d_d	Space-Time Dimension r
	Superverse Space			
0	Complex Octonion Octonion Octonion (1024)	Complex Octonion Octonion Octonion	1024×1024	18
1	Octonion Octonion Octonion (512)	Octonion Octonion Octonion	512×512	16
2	Quaternion Octonion Octonion (256)	Quaternion Octonion Octonion	256×256	14
3	Complex Octonion Octonion (128)	Complex Octonion Octonion	128×128	12
4	Octonion Octonion (64)	Octonion Octonion	64×64	10
	Maxiverse Space			
5	Quaternion Octonion (32)	Quaternion Octonion	32×32	8
	Megaverses Space			
6	Complex Octonion (16)	Complex Octonion	16×16	4
	Universe Space			
	Minispaces			
7	Quaternion (4)	Quaternion	4×4	4
8	Real (4)	Real (4)	4×4	4
9	Real (4)	Real (4)	4×4	4

Figure 8.2. The spectrum of the ten octonion spaces. The spaces are numbered from 0 through 9. The numbers in parentheses in column 2 are the number of array row/column dimensions. The items in column 3 are the number of rows of dimensions (1024, 512, 256, 128, 64, 32, 16, 4, 4, 4). Spaces 7, 8, and 9 are for the case of fermion-antifermion annihilation in a 4 dimension universe (6 dimensions minus 2 compact dimensions).

9. Symmetry Splitting in Octonion Cosmology Based on Fermion-Antifermion Annihilation

This chapter shows the origin of symmetry splitting due to the chain of fermion-antifermion annihilations, and particularly due to the structure of the dimension arrays derived from spinor arrays.

9.1 General Pattern of Splittings due to Cayley Form of Spectrum

Cayley numbers are generated as powers of the integer two. Spinor Cayley number arrays are necessarily square. So octonion spectrum dimension arrays increase in size by factors of four. Thus the Cayley-Dickson number $n = 9$ space can be viewed as four copies of the $n = 8$ space. Their connection via fermion-antifermion annihilation indicates they are copies.

Iterating this process down the chain of annihilations to space creations leads to a composite view of dimension space 9 as a nested set of dimension arrays of the lower spaces. See Fig. 9.1.

The pattern of quadrisections gives the following partition of the $n = 9$ dimension array into dimension sub-blocks, some of which directly appear in NEWQUeST and the New Unified SuperStandard Theory (NEWUST):

Array	Number of Components	QUeST Blocks
512 × 512	262,144	
256 × 256	65,536	
128 × 128	16,384	
64 × 64	4096	
32 × 32	1024	size of Megaverse Array
16 × 16	256	size of Universe Array
8 × 8	64	size of one layer
4 × 4	16	size of $SU(2) \otimes U(1) \otimes SU(4) \otimes SL(2, \mathbf{C})$[19] size of $U(4) \otimes U(4)$
2 × 2	4	size of $SU(2) \otimes U(1)$ size of $SL(2, \mathbf{C})$

The symmetry structure of NEWQUeST and NEWUTMOST embody the structure of the chain of octonion spaces in chapter 7. The symmetry group structure of the octonion spaces is directly connected to the structure of the spinor arrays within the scalar space particles created by fermion-antifermion annihilation. The following section shows the connection in detail for the NEWUTMOST dimension array for the Megaverse.

---

[19] The Lorentz group. $SO^{+}(1,3)$ in particular

The basis of the observed symmetries in the fermion-antifermion annihilation model of Octonion Cosmology raises the possibility that "symmetry breakdown" does not apply to the observed symmetries except, perhaps, for symmetry breakdown in the ElectroWeak SU(2)⊗U(1) sectors.

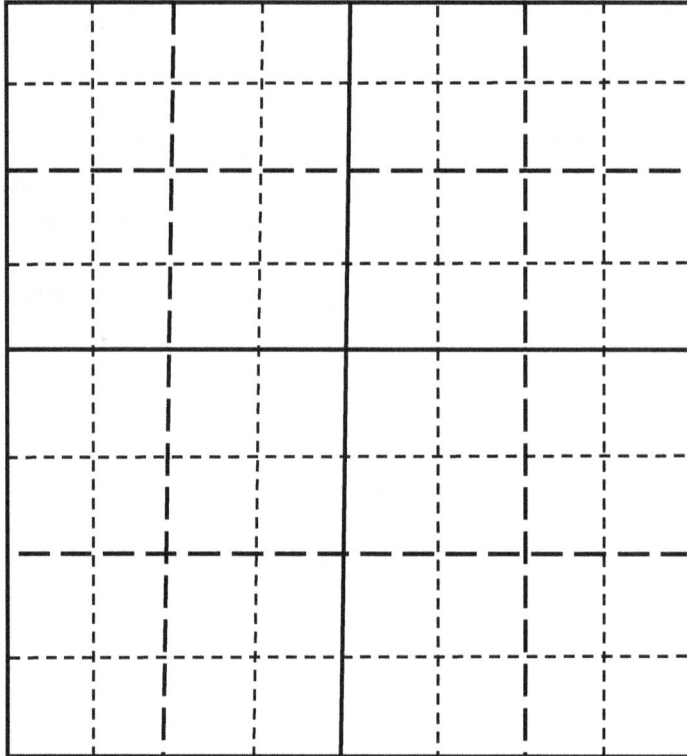

Figure 9.1. The Cayley-Dickson number n = 9 dimension array quadrisected three times to n = 7 dimension arrays. Quadrisection continues down to n = 1 for the spectrum of Fig. 8.1, and down to n = 4 (spectrum number 6) for the spectrum of Fig. 8.2. There are 64 n = 7 space parts visible above.

9.2 Detailed Form of Splitting Based on Spinor Array Structure

The fermion-antifermion annihilation process was described in chapter 6. A fermion-antifermion pair with momenta p_+ and p_- generated a scalar space particle with momentum $p_s = p_+ + p_-$.

$$U(p_-)\ "x"\ V(p_+) \rightarrow U(p_+ + p_-) \rightarrow DimensionArray \rightarrow Set\ of\ Symmetries$$

The spinor array of the created particle yields a dimension array based on the $U(2^{r/2})$ symmetry. We now turn to consider the detailed form of the space-time spinor array and show it reveals the symmetry of NEWQUeST and NEWUST.

The breakup of the spinor array into sets of large and small components is identical to the splitting of the dimension array into sets of dimensions of internal symmetries and sets of fermions excepting ElectroWeak symmetry breakdowns. Compare Fig. 9.2a with Figs. 9.2b and 9.2c.

9.2.0 Symmetry Splitting Based on Spinor Array Structure[20]

There is a general belief that there exists an overall symmetry for an entire universe/Megaverse. Spontaneous symmetry breakdowns are viewed as the source of the separation of particle symmetries into SU(3), SU(2)⊗U(1), and so on, as well as the breaking of symmetries such as ElectroWeak SU(2)⊗U(1) breakdown.

We put forward the proposition that the presumed inherent U(128) symmetry of the 256 dimension NEWQUeST universes (and similarly for Megaverses and other spaces) never existed.[21] Universes and Megaverses *began* with factored symmetries from their moment of origin in a fermion-antifermion annihilation.

We have described the splits generating the 10 octonion spaces, which we also believe existed from the first "moment" of existence of the Cosmos.

We see three types of symmetry splitting:

1. Splitting of the Superverse into 10 octonion spaces. We call this *global splitting by inheritance* since it generates entire octonion spaces.

2. Splitting of each octonion space into sets of Dimension-32 atoms. We call this type of splitting *local splitting by inheritance* since it splits the symmetry of an individual octonion space.

3. Splitting of a symmetry within a Dimension-32 atom such as SU(2)⊗U(1), SU(4), a U(4) Generation group, and a U(4) Layer group. This splitting is accomplished by spontaneous symmetry breaking.[22]

---

[20] This section previously appeared in Blaha (2021b).

[21] Chapter 7 suggests the Superverse is the origin of the ten octonion spaces. The ten spaces are shown as a "splitting" of the Superverse. As above, we propose that the split-generated factoring of spaces existed from the "Beginning" – not as a result of spontaneous symmetry breaking.

[22] At best, splitting of type 3 is the only splitting that may be relevant to running coupling constant estimates of the symmetry unification energy.

This section describes splitting of type 2. Spontaneous symmetry breaking (type 3) is described in Blaha (2018e) and (2020c).

9.2.1 Splitting of Type 2 Local Splitting by Inheritance

We saw in Blaha (2021a) that the overall structure of particle symmetries is determined in the large by the spinor structure of the annihilating fermion-antifermion pairs that generate universes[23] and megaverses.[24] *Thus there is a splitting of symmetries into a product of factors that is not due to spontaneous symmetry breaking (as it is usually envisioned.[25]) We call this type of splitting local splitting by inheritance since it is, in a very real sense, inherited from a "parent" octonion space instance.*

9.2.2 NEWQUeST Inheritance

The spinor structure of an annihilating fermion-antifermion pair in a NEWUTMOST Megaverse causes 4×4 blocks of dimensions in the resulting NEWQUeST universe dimension array. It factors the overall symmetry into blocks of internal symmetries. The blocks are not the result of symmetry breaking but reflect the structure of the spinors in NEWUTMOST Megaverse particles.[26] The sixteen 16-spinors of an 8-dimension Megaverse scalar or fermion are depicted in Figs. 9.2 and 9.3.: Fig. 9.4 shows the resulting internal symmetry array 4×4 block structure of NEWQUeST.

9.2.3 NEWUTMOST Megaverse Inheritance

The spinor structure of an annihilating fermion-antifermion pair in the Maxiverse causes 8×8 blocks of dimensions in the resulting NEWUTMOST Megaverse that determine blocks of internal symmetries.[27] These blocks (and the corresponding blocks for NEWQUeST) are described in detail in chapter 8 of Blaha (2021a). The blocks are not the result of symmetry breaking but reflect the structure of the spinors in Maxiverse urfermions (fermions). The thirty-two 32-spinors of a 10 dimension Maxiverse urfermion are depicted in Fig. 9.5. Spinor parts map to 8×8 blocks of internal symmetries (Fig. 9.6) in the NEWUTMOST Megaverse.

9.2.4 NEWQUeST – NEWUST Map and Map to NEWUTMOST

The 4×4 blocks of NEWQUeST are naturally determined by a map from NEWQUeST to the New Unified SuperStandard Theory (NEWUST) that is shown in chapter 3 of Blaha (2019a). *The composition of the 8×8 blocks in the NEWUTMOST*

---

[23] Chapter 4 of Blaha (2021a).

[24] Chapter 8 of Blaha (2021a).

[25] Spontaneous symmetry breaking does appear to take place in ElectroWeak Theory: SU(2)⊗U(1) breaking, and in the Generation and Layer groups, as well as the breaking of Strong groups from SU(4) to SU(3)⊗U(1).

[26] For that reason this form of symmetry factoring can be viewed as evidence for the existence of our universe in a Megaverse.

[27] The breakdown into 64 dimension blocks carries over to NEWQUeST. Each layer of NEWQUeST is a 64 dimension block.

Megaverse is determined by mapping upward from NEWQUeST since the NEWUTMOST dimension array consists of four copies of NEWQUeST.

Number of Columns = 4   4   4   4

u-type fermion)		v-type (anti-fermion)	
4 u spin up		v small terms 1	
4	u spin down	v small terms 2	
4 u small terms 1		v spin down	
4 u small terms 2			v spin up

Figure 9.2a. The 16-spinors of a 8 space-time dimension (a Megaverse) spinor array. Each spinor column has 16 rows. There are 16 16-spinors.

SU(4) Fermion Particle Periodic Table

Number of Columns = 4   **NORMAL**   4   4   **DARK**   4

	NORMAL		DARK	
Layer 1 4	e 3 up-quarks u spin up	v 3 down-quarks	e 3 up-quarks v small terms 1	v 3 down-quarks
Layer 2 4	e 3 up-quarks	v 3 down-quarks u spin down	e 3 up-quarks v small terms 2	v 3 down-quarks
Layer 3 4	e 3 up-quarks u small terms 1	v 3 down-quarks	e 3 up-quarks v spin down	v 3 down-quarks
Layer 4 4	e 3 up-quarks u small terms 2	v 3 down-quarks	e 3 up-quarks	v 3 down-quarks v spin up

Figure 9.2b. The 16-spinors of a 8 space-time dimension spinor array. It becomes the dimension array of a universe. It corresponds directly with SU(4) **4** (or SU(3)⊗U(1)) fermions that are shown superimposed on the array. Each spinor column has 16 rows. There are 16 16-spinors. There are four layers. Each set of 4 fermions has 4 generations matching the number of rows in each

layer. The Periodic Table is broken into Normal and Dark sectors. As shown above a Match!

NEWQUeST/NEWUST Symmetries Table

Number of Columns =	4	**NORMAL**	4		4	**DARK**	4
Layer 1	SU(4)⊗ Generation U(4)	SU(2)⊗U(1)⊗SL(2,C)[28] Layer U(4)			SU(4)⊗ Generation U(4)	SU(2)⊗U(1)⊗SL(2,C) Layer U(4)	
4		u spin up				v small terms 1	
Layer 2	SU(4)⊗ Generation U(4)	SU(2)⊗U(1)⊗SL(2,C) Layer U(4)			SU(4)⊗ Generation U(4)	SU(2)⊗U(1)⊗SL(2,C) Layer U(4)	
4		u spin down				v small terms 2	
Layer 3	SU(4)⊗ Generation U(4)	SU(2)⊗U(1)⊗SL(2,C) Layer U(4)			SU(4)⊗ Generation U(4)	SU(2)⊗U(1)⊗SL(2,C) Layer U(4)	
4		u small terms 1				v spin down	
Layer 4	SU(4)⊗ Generation U(4)	SU(2)⊗U(1)⊗SL(2,C) Layer U(4)			SU(4)⊗ Generation U(4)	SU(2)⊗U(1)⊗SL(2,C) Layer U(4)	
4		u small terms 2				v spin up	

Figure 9.2c. The16-spinors of a 8 space-time dimension spinor array. It becomes the dimension array of a universe. It corresponds directly with the pattern of symmetries of NEWUST. Note each layer has its own set of internal symmetries. Normal and Dark sector symmetries are also different. The Layer Groups and the Connection Groups tie the various parts together. As shown, the pattern of symmetries closely matches the NEWQUeST dimension array (spinor array) splittings.The 4 ×4 blocks match the symmetry splits SU(4) appears split into SU(3)⊗U(1) experimentally.The eight SL(2, **C**) groups are transformed into one 4-dimension space-time and seven U(2) connection groups. See Fig. 11.3 and section 11-A.4 for details

---

[28] The Lorentz group. SO$^+$(1 ,3) in particular

u-up–v-v1	u-up–v-v2	u-up–v-down	u-up–v-up
u-up–v-v1	u-up–v-v2	u-up–v-down	u-up–v-up
u-up–v-v1	u-up–v-v2	u-up–v-down	u-up–v-up
u-up–v-v1	u-up–v-v2	u-up–v-down	u-up–v-up
u-down–v-v1	u-down–v-v2	u-down–v-down	u-down–v-up
u-down–v-v1	u-down–v-v2	u-down–v-down	u-down–v-up
u-down–v-v1	u-down–v-v2	u-down–v-down	u-down–v-up
u-down–v-v1	u-down–v-v2	u-down–v-down	u-down–v-up
u-v1–v-v1	u-v1–v-v2	u-v1–v-down	u-v1–v-up
u-v1–v-v1	u-v1–v-v2	u-v1–v-down	u-v1–v-up
u-v1–v-v1	u-v1–v-v2	u-v1–v-down	u-v1–v-up
u-v1–v-v1	u-v1–v-v2	u-v1–v-down	u-v1–v-up
u-v2–v-v1	u-v2–v-v2	u-v2–v-down	u-v2–v-up
u-v2–v-v1	u-v2–v-v2	u-v2–v-down	u-v2–v-up
u-v2–v-v1	u-v2–v-v2	u-v2–v-down	u-v2–v-up
u-v2–v-v1	u-v2–v-v2	u-v2–v-down	u-v2–v-up

Figure 9.3. The product array [U_{ba}] of the composite u-type and v-type spinors illustrating the structure of the product array of u's and v's. The result of fermion-antifermion annihilation as indicated in chapter 6.

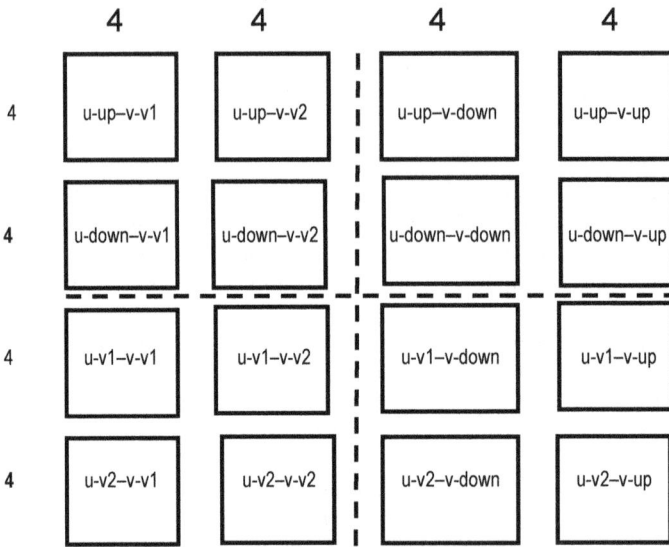

Figure 9.4. Block form of the 16 × 16 [U_{ba}] array. This is also the form of the NEWQUeST dimension array of 256 dimensions. The blocks are divided by dashed lines that separate 64 dimension sections. These sections map to layers in NEWQUeST (NEWUST).

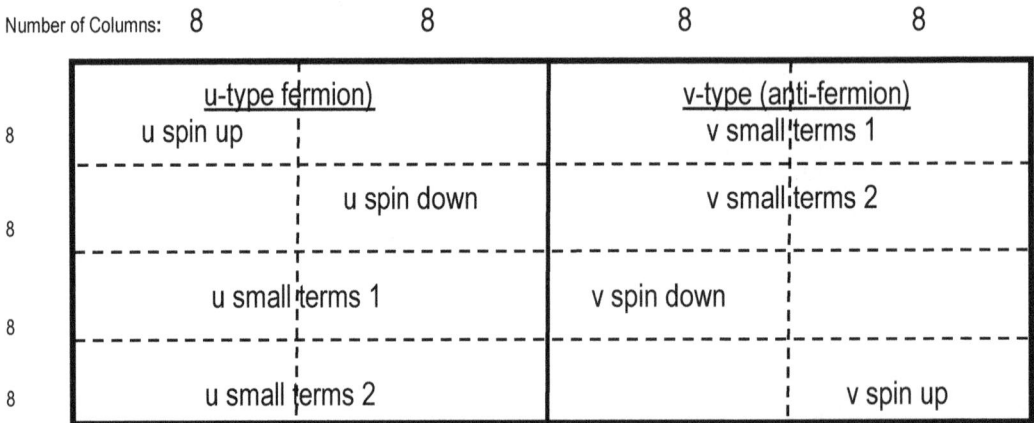

Figure 9.5. The thirty-two 32-spinors of a 10 dimension urfermion (an n = 6 Maxiverse fermion or scalar). Each spinor column has 32 rows.

u-up–v-v1	u-up–v-v2	u-up–v-down	u-up–v-up
u-up–v-v1	u-up–v-v2	u-up–v-down	u-up–v-up
u-up–v-v1	u-up–v-v2	u-up–v-down	u-up–v-up
u-up–v-v1	u-up–v-v2	u-up–v-down	u-up–v-up
u-up–v-v1	u-up–v-v2	u-up–v-down	u-up–v-up
u-up–v-v1	u-up–v-v2	u-up–v-down	u-up–v-up
u-up–v-v1	u-up–v-v2	u-up–v-down	u-up–v-up
u-up–v-v1	u-up–v-v2	u-up–v-down	u-up–v-up
u-down–v-v1	u-down–v-v2	u-down–v-down	u-down–v-up
u-down–v-v1	u-down–v-v2	u-down–v-down	u-down–v-up
u-down–v-v1	u-down–v-v2	u-down–v-down	u-down–v-up
u-down–v-v1	u-down–v-v2	u-down–v-down	u-down–v-up
u-down–v-v1	u-down–v-v2	u-down–v-down	u-down–v-up
u-down–v-v1	u-down–v-v2	u-down–v-down	u-down–v-up
u-down–v-v1	u-down–v-v2	u-down–v-down	u-down–v-up
u-down–v-v1	u-down–v-v2	u-down–v-down	u-down–v-up
u-v1–v-v1	u-v1–v-v2	u-v1–v-down	u-v1–v-up
u-v1–v-v1	u-v1–v-v2	u-v1–v-down	u-v1–v-up
u-v1–v-v1	u-v1–v-v2	u-v1–v-down	u-v1–v-up
u-v1–v-v1	u-v1–v-v2	u-v1–v-down	u-v1–v-up
u-v1–v-v1	u-v1–v-v2	u-v1–v-down	u-v1–v-up
u-v1–v-v1	u-v1–v-v2	u-v1–v-down	u-v1–v-up
u-v1–v-v1	u-v1–v-v2	u-v1–v-down	u-v1–v-up
u-v1–v-v1	u-v1–v-v2	u-v1–v-down	u-v1–v-up
u-v2–v-v1	u-v2–v-v2	u-v2–v-down	u-v2–v-up
u-v2–v-v1	u-v2–v-v2	u-v2–v-down	u-v2–v-up
u-v2–v-v1	u-v2–v-v2	u-v2–v-down	u-v2–v-up
u-v2–v-v1	u-v2–v-v2	u-v2–v-down	u-v2–v-up
u-v2–v-v1	u-v2–v-v2	u-v2–v-down	u-v2–v-up
u-v2–v-v1	u-v2–v-v2	u-v2–v-down	u-v2–v-up
u-v2–v-v1	u-v2–v-v2	u-v2–v-down	u-v2–v-up
u-v2–v-v1	u-v2–v-v2	u-v2–v-down	u-v2–v-up

Figure 9.6. The 32×32 product array (an n = 5 Megaverse) of the composite u-type and v-type spinors illustrating the structure of the product array of u's and v's. Each column represents 8 spinor columns, making 32 columns in all. The result of fermion-antifermion annihilation as indicated in chapter 6.

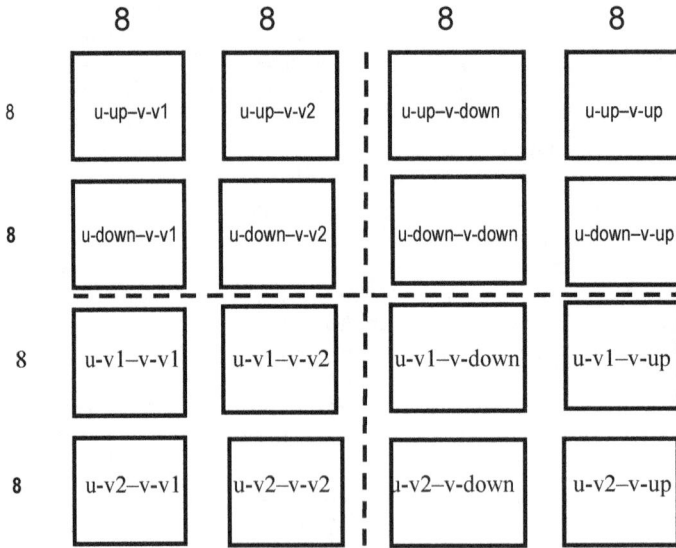

Figure 9.7. Block form of a 32 × 32 array. This is also the form of the NEWUTMOST dimension array of 1024 dimensions. It has 16 × 16 blocks, each of which have four 8 × 8 blocks within them. The four sets of 16 × 16 blocks above are divided by dashed lines. They form 256 dimension sections. These 256 dimension sections map to layers in NEWUTMOST. Each NEWUTMOST layer is equivalent to a NEWQUeST 256 dimension array.

10. Dressing a Φ Space Particle

The Φ space particle defined in chapter 2 has a $U(2^{r/2})$ unitary group, which determines its internal symmetry and space-time groups.

However it does not appear to create universes, multiverses, and so on, containing the usual accouterments of particles and of energies of physical spaces. A dimension array, by itself, is not a universe, multiverse, and so on. We must "dress" each space with particles and energy.

The situation is not unlike C++ where one defines a class specifying variables within it, and then creates objects of the class type by allocating chunks of memory.

We turn now to the analogous dressing of spaces with symmetries, energy and particles.

10.1 Symmetry Dressing

The dimension array resulting from a spinor array created from a fermion-antifermion annihilation into a scalar particle enables a set of symmetries to be allocated by splitting dimensions into sets of fundamental representation dimensions. This conceptual process requires no work. But it creates a framework for the specification of the elementary particle spectrum.

10.2 Energy/Particle Dressing

Each fundamental symmetry has a set of fundamental fermions in a fundamental representation. Each symmetry also has associated sets of gauge vector bosons and Higgs bosons. To those particles we add, at minimum, Gravitation and its gravitons.

The generation of the particles requires energy to be present. The mass m' of the created scalar boson is the likely source of the energy of the created instance of the space. Thus the source of the energy of a child space instance is in the parent space instance.[29]

It is reasonable to assume that the mass-energy is initially concentrated at the creation point. It is also reasonable to assume the energy embodied in the particles has a Planck black body distribution.[30] The expansion of the space follows.

10.3 A View of Creation

We have seen the progression from the new Quantum Space Theory to Octonion Cosmology with a fairly detailed relation between the spinor array generated by an annihilating fermion-antifermion pair and symmetries. This achievement raises the possibility that Quantum Field Theory, upon which Quantum Space Theory is based, and which originated in the study of electrons, protons and neutrons, may be the true

---

[29] The energy of the top space (God Space) is assumed. All energies of all child spaces flow from it.
[30] Blaha (2004) and (2021d).

"language" of Nature.[31] It also raises the possibility that a more fundamental substratum of Reality exists, from which Quantum Field Theory proceeds.

---

[31] The similarity of NEWQUeST structure and DNA structure also raises the possibility of a stability principle for structures that applies to both DNA and Octonion Cosmology. See Blaha (2021e).

11.   NEWQUeST and NEWUST

11.0 Brief Description

Originally the author developed a theory of elementary particles called the Unified SuperStandard Theory (UST) in the years preceding 2020. Starting in January, 2020 the author developed the QUeST quantum field theory based on quaternions and octonions, which, to the author's surprise, had the same set of symmetries as UST including the Generation a

nd Layer groups. In the past 1.75 years the author modified[32] UST and QUeST by "tradimg" internal symmetry coordinates for space-time coordinates. The results were NEWUST and NEWQUeST and a Megaverse space NEWUTMOST. This chapter and appendix describes these theories.

11.1 Detailed Summary of UST, QUeST, NEWUST, and NEWQUeST

This section summarizes features of these four equivalent theories. The theories can be found in

Blaha, 2018e, *Unification of God Theory and Unified SuperStandard Model THIRD EDITION*

Blaha, 2020a, *Quaternion Unified SuperStandard Theory (The QUeST) and Megaverse Octonion SuperStandard Theory (MOST)*

Blaha, 2020c, *Unified SuperStandard Theories for Quaternion Universes & The Octonion Megaverse*

Blaha, 2021c, *Beyond Octonion Cosmology*

Blaha, 2021d, *Universes are Particles*

Blaha, 2021e, *Octonion-like dna-based life, Universe expansion is decay, Emerging New Physics*

Blaha, 2021f, *The Science of Creation New Quantum Field Theory of Spaces*

as well as earlier books by the author.

The theories originated in the past twenty years from the Standard Model of Particles with SU(2)⊗U(1)⊗SU(3), and Two-Tier Quantum and PseudoQuantum Field Theory. Noting the presence of conserved particle numbers, and the presence of at least three fermion generations, we introduced the U(4) Generation Groups and the U(4) Layer Groups together with four layers of four generation fermions. The "Normal" fermions had a corresponding set of four layers of four generations of "Dark" fermions. The result was the Unified SuperStandard Theory (UST) symmetry:

---

[32] Blaha (2021b) and (2021c).

$$\{[SU(2)\otimes U(1)\otimes SU(3)]^2\otimes U(4)^4\}^4$$

with an additional Strong Interaction U(1) group (analogous to that of ElectroWeak theory) found to be needed. Space-time had four dimensions.

In January, 2020 the author discovered that an octonion-based theory that he constructed and named QUeST had the same internal symmetries as UST with the addition of $U(1)^8$. QUeST's internal symmetry, which could be based on a 16×16 dimension array, was

$$[SU(2)\otimes U(1)\otimes SU(3)\otimes U(1)]^8\otimes U(4)^{16}$$

The addition of $U(1)^8$ indicates that the Strong Interactions in the theory are broken Strong SU(4).

During 2020 the author developed an octonion space spectrum for both universes and Megaverses, and other spaces. The spectrum was shown to arise from a generation mechanism whereby fermion-antifermion annihilation in a higher space produced an instance of a lower space. A critical part of the derivation of the octonion spectrum was the realization that even space-time dimension spinor arrays are composed of Cayley number rows and columns. Spinor arrays of annihilating fermion-antifermion pairs were shown to generate the dimension arrays of subspace instances. Analyzing the spinor arrays the author noted that the dimension array could be viewed as composed of 64 dimension blocks, which were further subdivided into 16 dimension subblocks.

The subblock structuring, using the known contents of the Standard Model plus Generation and Layer groups, gives the dimension array structure containing 4×4 subblocks in Figs. 11.1 and 11.2.

Thus there was a *most* satisfactory match between UST and QUeST with the only significant difference being the space-time: four octonion (complex quaternion) coordinates for QUeST and four real space-time coordinates for UST.

The form of the square spinor arrays generated by fermion-antifermion annihilation gives 64 dimension blocks and 16 dimension blocks as well as 32 dimension composite blocks that are evidenced in the NEWQUeST (and NEWUTMOST) fermion spectrums and internal symmetry group structure.

Since we see only real dimensions in Reality, we recently transferred 28 QUeST dimensions from space-time to $U(2)^7$ internal symmetry dimensions. The set of internal symmetries was increased by $U(2)^7$, which we call Connection Groups.[33] Each Connection group specifies interactions between corresponding fermions (e with e, q with q, and so on) in separate layers and between Normal and Dark fermions. The connections between the various blocks of fermions is shown in Fig. 11.3. *We implement the very practical rule that all blocks must be connected by interactions or they would not be of physical interest. A totally isolated block effectively does not exist physically (except possibly for gravitation effects).*

---

[33] See section 11-A.4.

The interactions of the Connection groups must be very weak and/or their gauge bosons must be very massive.

The addition of the Connection Groups and the reduction of space-time dimensions accordingly results in NEWQUeST and NEWUST as summarized below. See appendix 11-A for a discussion of NEWQUeST groups.

Note: the Generaation, Layer, and Connection groups are all badly broken. Their vector bosons must be very massive since they have not been detected in experiments.

11.1.1 Internal Symmetries

The groups are ElectroWeak SU(2)⊗U(1), Strong SU(3), Generation Group U(4), Layer Group U(4), and U(2) and U(4) Connection groups obtained by transfer from space-time coordinates (See Blaha 2012c). The SU(3)⊗U(1) symmetry may be a broken SU(4) symmetry in eqs. 11.1 – 11.4. The internal symmetries for the theories are:

UST

$$[SU(2) \otimes U(1) \otimes SU(3)]^8 \otimes U(4)^{16} \tag{11.1}$$

QUeST

$$[SU(2) \otimes U(1) \otimes SU(3) \otimes U(1)]^8 \otimes U(4)^{16} \tag{11.2}$$

NEWQUeST

$$[SU(2) \otimes U(1) \otimes SU(3) \otimes U(1)]^8 \otimes U(4)^{16} \otimes U(2)^7 \tag{11.3}$$

The only change is in internal symmetries: Twenty-eight real dimensions transferred from space-time coordinates to Connection group $U(2)^7$ internal symmetry.

NEWUST

$$[SU(2) \otimes U(1) \otimes SU(3) \otimes U(1)]^8 \otimes U(4)^{16} \otimes U(2)^7 \tag{11.4}$$

The only change in internal symmetries: Twenty-eight real dimensions added for $U(2)^7$ Connection group internal symmetry.

11.1.2 Space-Time Coordinates
UST

Four real space-time coordinates.

QUeST

Four octonion (complex quaternion) coordinates.

NEWUST

Four real space-time coordinates. No change from UST space-time.

NEWQUeST

Four real space-time coordinates. The six coordinates in the $n = 4$ octonion space (Figs. 1.5 and 1.6) were lowered to four space-time coordinates with two coordinates transferred to Connection groups.

The only change is in space-time coordinates: Fourteen dimensions transferred from QUeST space-time coordinates to Connection group $U(2)^7$ internal symmetry.

11.2 Fundamental Fermion Spectrum

There are 256 fundamental fermions[34] in NEWQUeST and NEWUST. Conceptually their structure can be viewed as an extrapolation of the known three generations of The Standard Model. For reasons given previously and in Appendix 11-A a fourth generation was indicated and a corresponding Dark sector of similar structure was added. In addition, because of the need for Layer groups (Appendix 11-A), the overall structure consisted of four copies of this layer.

Correspondingly, each layer also has its own set of internal symmetry gauge groups[35] to limit mixing between the layers to Layer group interactions and Connection group interactions.

Fig. 11.4 shows the structure of the NEWQUeST/NEWUST fermions. The blocks are 4×4 reflecting the origin of the NEWQUeST/NEWUST space (universe) instance from Megaverse fermion-antifermion annihilation. The spinor analysis of their spinor arrays yields a 16 dimension-based block structure. The 64 dimension fermion layers reflect the 64 dimension structuring of the Megaverse obtained from its creation by fermion-antifermion creation in the Maxiverse. . .

11.3 Total Dimensions

The total of internal symmetry and space-time dimensions is 256 in all four theories listed above. It is based on the 16×16 dimension array of the Cayley number n = 4 space of the Octonion spectrum.

11.4 Pattern of Internal Symmetries

The NEWQUeST dimension array for internal symmetries is subdivided into four layers of 56 dimensions—just as in NEWUST (and UST). Fig. 11.1 displays the layers using SU(4) in place of SU(3)⊗U(1). Each layer has a block of 28 dimensions for Normal and 28 dimensions for Dark sectors. There are also seven U(2) Connection groups[36] plus four real-valued space-time coordinates bringing the NEWQUeST total to 256 dimensions.= 4*56 + 28 + 4 = 256 dimensions. The Connection groups are shown in Fig. 11.2 (and section 11-A.4), which is a revision of the pattern shown in Blaha (2021c).

---

[34] See Fig. 9.2b for the fermion spectrum determined by a spinor array.
[35] See Fig. 9.2c for the pattern of symmetry groups.
[36] See Appendix 11-A.

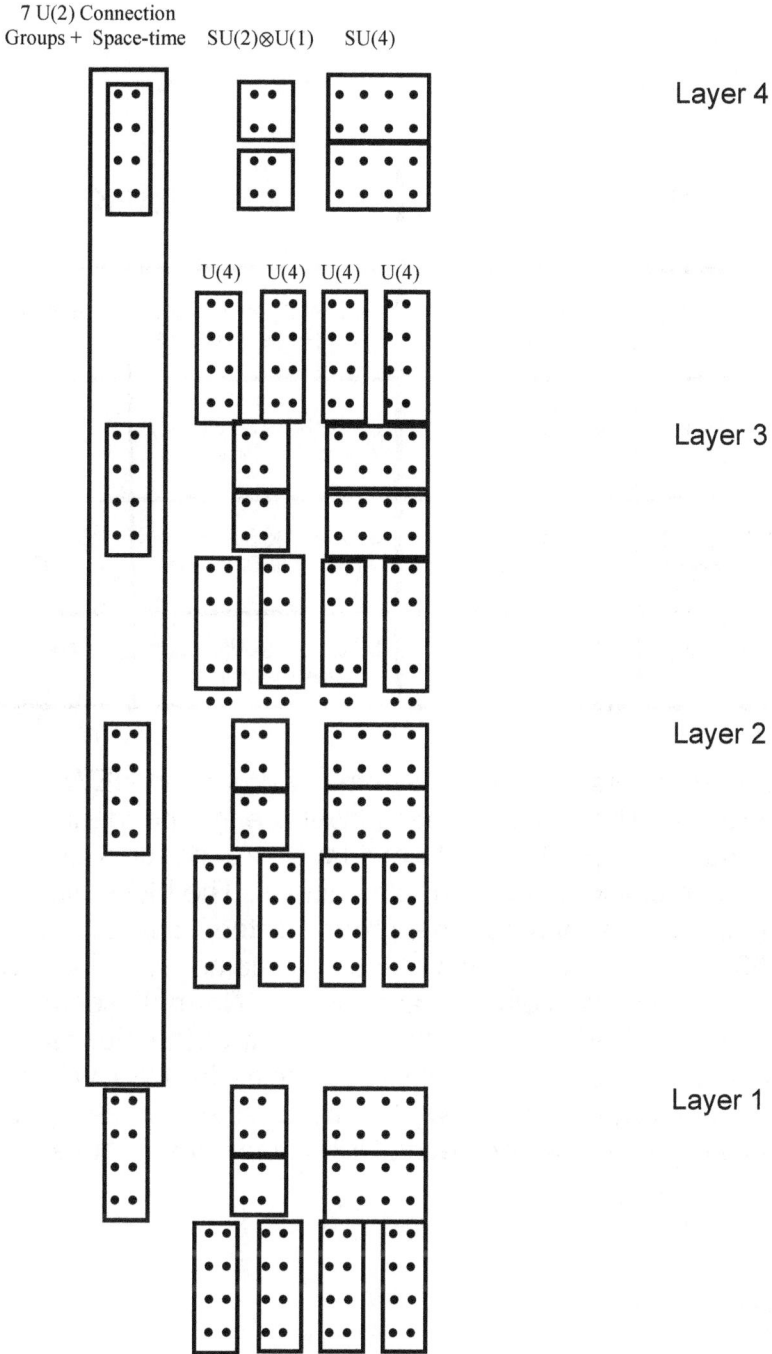

Figure 11.1. The four layers of NEWUST and NEWQUeST internal symmetry groups (and space-time) with SU(4) before breakdown to SU(3)⊗U(1). Note the left column of blocks combine to specify a 4 dimension real space-time plus seven U(2) Connection groups. Note each layer has 64 dimensions = 56 + 8 dimensions.

Layers ↓	NORMAL		DARK	
	4	4	4	4
4	SU(2)⊗U(1)⊗SU(3)⊗U(1) 4 Space-time Dimensions	Generation + Layer Groups	SU(2)⊗U(1)⊗SU(3)⊗U(1) 4 Space-time Dimensions	Generation + Layer Groups
4	SU(2)⊗U(1)⊗SU(3)⊗U(1) 4 Space-time Dimensions	Generation + Layer Groups	SU(2)⊗U(1)⊗SU(3)⊗U(1) 4 Space-time Dimensions	Generation + Layer Groups
4	SU(2)⊗U(1)⊗SU(3)⊗U(1) 4 Space-time Dimensions	Generation + Layer Groups	SU(2)⊗U(1)⊗SU(3)⊗U(1) 4 Space-time Dimensions	Generation + Layer Groups
4	SU(2)⊗U(1)⊗SU(3)⊗U(1) 4 Space-time Dimensions	Generation + Layer Groups	SU(2)⊗U(1)⊗SU(3)⊗U(1) 4 Space-time Dimensions	Generation + Layer Groups

Figure 11.2.. Four layers of Internal Symmetry groups in NEWQUeST (omitting Connection groups). The groups in each layer are independent of those in other layers. The groups in each block of each layer are independent of those in the other blocks. Each block contains 16 dimensions. The block dimensions furnish fundamental representations for the groups listed. The entire set of blocks contains 256 dimensions. Each layer contains 56 internal symmetry dimensions. The first two columns are for the "Normal" sector. The last two columns are for the "Dark" sector (although most of the Normal sector is Dark observationally at present.) This figure also holds for UST with the addition of U(1) groups. The eight sets of 4 real dimension space-times combine to give a 4 real dimension space-time and seven U(2) Connection groups.

NEWQUeST Vector Bosons

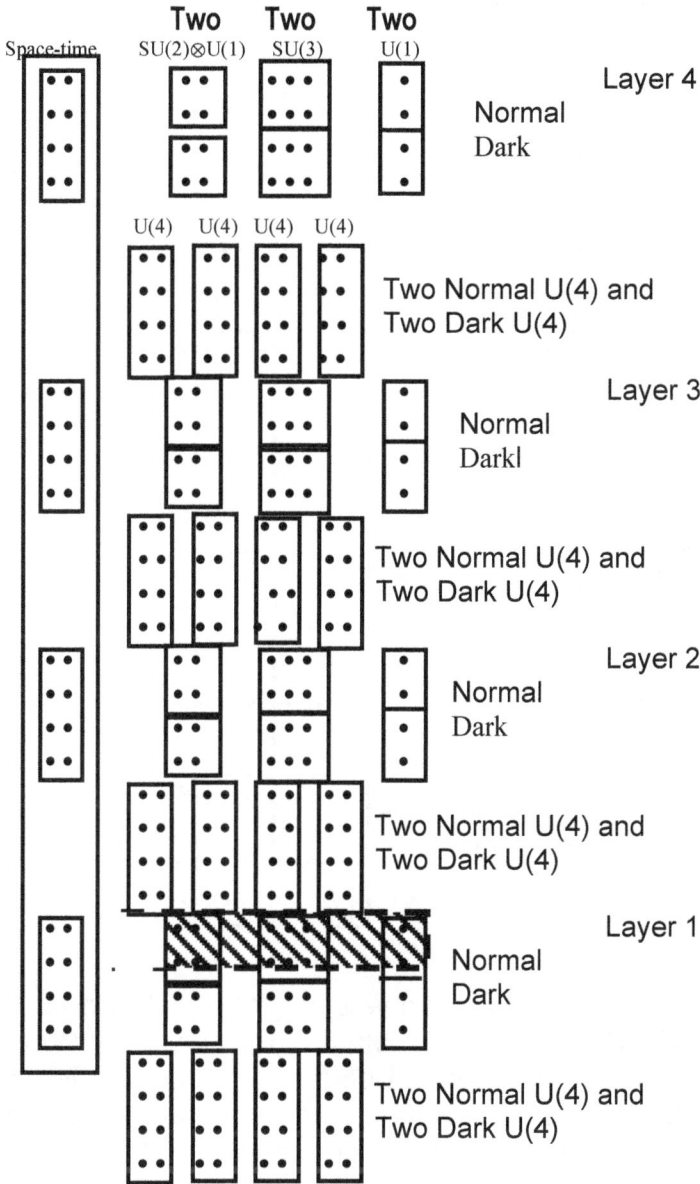

Figure 11.2a. The four layers of QUeST internal symmetry groups (and space-time) for the 32 octonion dimension form of space. Note: each row has an 8 • octonion. Note the left column of blocks combine to specify a 4 dimension octonion space-time and seven U(2) Connection Groups. Note each layer requires 64 dimensions.. *Note the duplication (using the "Two" label) of each symmetry in each layer. The cross hatched area indicates the known symmetry groups.*

Connection Group Applied to Fermions

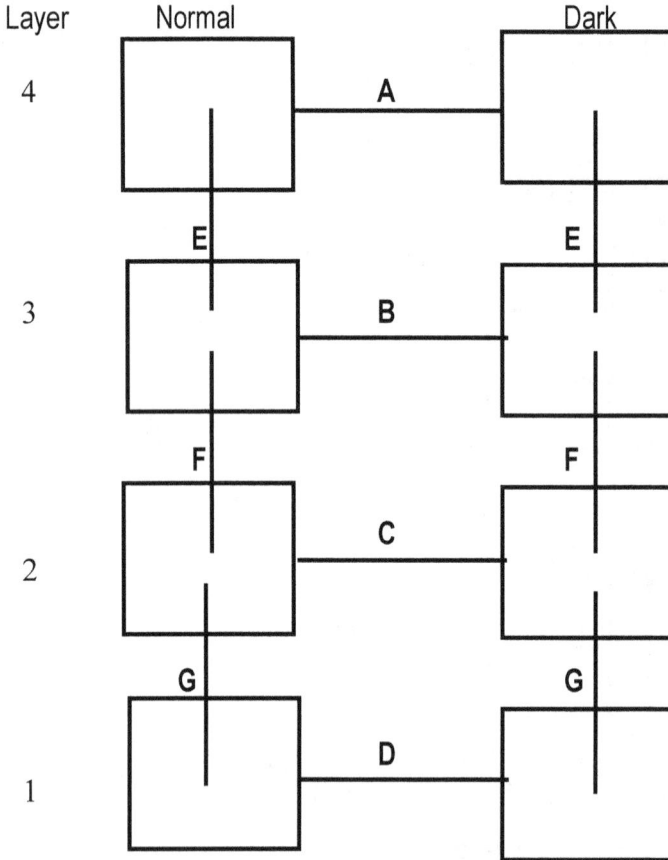

Figure 11.3. The seven U(2) Connection groups (shown as 10 lines) between the eight NEWQUeST/NEWUST blocks. Connection groups are obtained by transfering 28 dimensions from QUeST space-time to internal symmetries with the consequent reduction of the space-time from four octonion (complex quaternion) coordinates to four real coordinates. The Connection groups generate rotations and interactions between corresponding fermions and vector bosons of each pair of blocks. See Appendix 11-A. The Normal and Dark sector U(2) vertical connections above (E, F, G) represent the same U(2) groups.

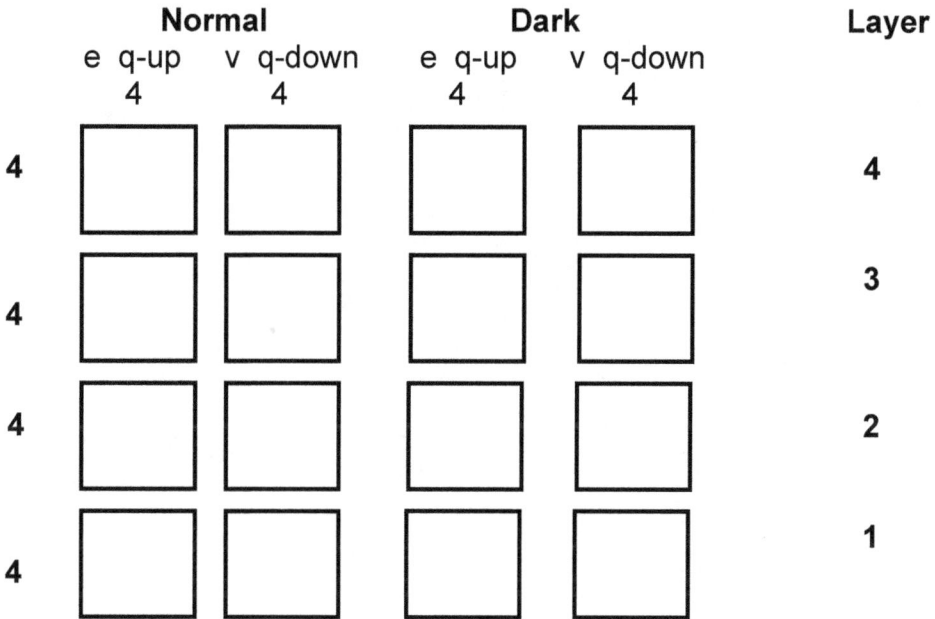

Figure 11.4. Block form of a 16 × 16 NEWQUeST/NEWUST fermion array with each block row corresponding to one layer. Each block contains four generations of fermions. The result is 4 × 4 blocks. The label e q-up indicates a charged lepton – up-type quark pair, v q-down indicates a neutral lepton – down-type quark pair, and so on. The blocks can be viewed as SU(3)⊗U(1) or broken SU(4) blocks.

Appendix 11-A. Description of the NEWQUeST/NEWUST Internal Symmetry Groups

NEWUST and NEWQUeST contain sets of symmetry groups: Strong SU(3), Generation group U(4), Layer group U(4) and Connection group U(2). In this Appendix we describe their features. See Blaha (2019e) or (2020c) for more details on the Generation and Layer groups.

11-A.1 Strong SU(3)

The Strong Interaction Groups separately support rotations and interactions among the three up-type quarks and of the three down-type quarks in each generation, each layer, and in the Normal and Dark sectors. Each layer in the Normal and Dark sectors have a separate SU(3) group. Thus NEWQUeST/NEWUST each have a total of eight SU(3) Strong Interaction groups as shown in Figs. 11.1 and 11.2.

The fermions in each layer of the Normal and Dark sectors are different. Thus there are eight sets of fermions in the two block pairs in Fig. 11.4.

11-A. 2 U(4) Generation Groups

In the Big Bang all particles were massless and all symmetries unbroken. Hence the four Normal particle number symmetries, and the four Dark particle number symmetries, are all "conserved" in the Big Bang. Afterwards conservation are then broken afterwards in most cases.

We define two particle number operators for normal up-quark particles and down-quark particles, B_{uq} and B_{dq}. Similarly we define two particle number operators for normal species "e" (electron) particles and species "v" particles, B_e and B_v. Similarly we define Dark matter equivalents:[37] B_{De}, B_{Dv}, B_{Duq}, and B_{Ddq}.

In the absence of symmetry breaking these fermion particle number operators would be conserved. Thus there are two sets of "diagonal" operators with associated U(4) groups for the Normal and Dark sectors. They are part of the Normal U(4) Generation Group and the Dark U(4) Generation Group.

The fermion fundamental representation of a U(4) group has four fermions. U(4) has rotations, and also interactions of the form $\overline{\Psi}\gamma\cdot B\cdot T\Psi$ where Ψ is a fermion four-vector, B is a 16 component U(4) gauge field, and T consists of 16 component 4×4 U(4) arrays.

In the case of the Generation Group the gauge fields have electric charge zero. Since the four species have different electric charges (1, 0, 2/3, -1/3) the U(4) gauge

---

[37] By analogy, we assume that there are four species of Dark matter: charged Dark leptons, neutral Dark leptons, Dark up-type quarks, and Dark down-type quarks. Thus we are led to the Dark particle numbers: Dark Baryon Numbers, and Dark Lepton Numbers shown above.

boson fields cannot mix the fermions of different species. Generation Group interactions are diagonal[38] in fermion species (e, ν, up-quark, and down-quark species).

Consequently the U(4) Generation Group must have a reducible representation D consisting of a set of four fundamental U(4) representations, D_e, D_v, D_{upq}, and D_{dnq}, appearing in blocks along the diagonal of D. Each block is a separate U(4) irreducible representation for a species (due to the electric charge superselection rule.)

There are four generations of each species in the Normal and in the Dark matter sectors. The four generations for each fermion species: e, ν, up-quark, and down-quark each furnish a U(4) fundamental representation within the reducible representation D. The fourth generation of normal fermions has not as yet been found due to their extremely large masses.

The Generation Group rotates the fundamental fermions of each fundamental representation separately for each of the four species of each of the four layers.[39] Thus the Generation Group guarantees that all generations of each species have the same electric charge and other quantum numbers.

The U(4) Generation Group also specifies a gauge field interaction among the fermions of its fundamental representation, species by species, for both Normal and Dark sectors. The form of the interactions for the Normal sector for each fermion layer is:

$$g_e\overline{\Psi}_e\gamma\cdot B_e\cdot T\Psi_e + g_v\overline{\Psi}_v\gamma\cdot B_v\cdot T\Psi_v + g_{upq}\overline{\Psi}_{upq}\gamma\cdot B_{upq}\cdot T\Psi_{upq} + g_{dnq}\overline{\Psi}_{dnq}\gamma\cdot B_{dnq}\cdot T\Psi_{dnq}$$
$$(11\text{-}A.1)$$

where g_e ... are coupling constants, the gauge vector fields are B_e ... , and the Ψ_e ... are 4-vectors of fermions of the four generations of each species in a layer.

The gauge vector bosons of the Generation Group have large masses. If the conservation of the fermion particle numbers is broken then we view it as a consequence of Generation Group symmetry breaking.

Generation Group rotations guarantee the internal quantum numbers of each generation of each species are the same since symmetry breakdown is not present at the instant of the Big Bang.

The above discussion applies similarly to the Dark sector. There are 8 Generation Groups in total in NEWQUeST/NEWUST. See Figs. 11.1 and 11.2.

11-A.3 U(4) Layer Groups

The set[40] of particle number operators can be extended if we take account of the fourfold fermion generations.

We can subdivide the above particle number sets into four additional particle numbers *per generation*. For the ith generation (of the four generations) we define

---

[38] ElectroWeak interactions can cross between species due to their charged gauge vector bosons.

[39] There are separate Generation groups for each layer.

[40] Here again, in the Big Bang all particles were massless and all symmetries unbroken. Hence particle numbers are "conserved" in the Big Bang. Conservation is then broken afterwards in most cases.

L_{ie} – The "e" species particle number for the ith generation

L_{iv} – The v species particle number for the ith generation

L_{iuq} – The up-quark species particle number for the ith generation

L_{idq} – The down-quark species particle number for the ith generation

L_{iDe} – The Dark "e" species particle number for the ith generation

L_{iDv} – The Dark v species particle number for the ith generation

L_{iDuq} – The Dark up-quark species particle number for the ith generation

L_{iDdq} – Dark down-quark species particle number for the ith generation

for each generation i = 1, 2, 3, 4. Individual fermions have positive L_{ia} = +1 values and antifermions have negative L_{ia} = –1 values for each species.

At this point we have a set of four particle number operators for each of four generations (i = 1, 2, 3, 4) of fermions in the Normal sector and similarly in the Dark sector. We then define a U(4) group framework for each set of particle numbers.

The only way to specify fundamental representations for each of the four sets in a sector is to assume there are four layers, with each layer having four generations, and with a fundamental U(4) representation defined for each generation composed of fermions from each layer. Thus there are four Layer Groups for each Normal and each Dark sector: a Layer Group for generation 1, a Layer Group for generation 2, and so on.

The Layer Groups are also "split" by species due to the electric charge superselection rule. Each Layer Group is diagonal in the four fermion species. All their gauge fields are electrically neutral.

Consequently each of the four U(4) Layer Groups in the Normal fermion sector has a reducible U(4) representation D_j for j = 1, 2, 3, 4. Each reducible representation is composed of four irreducible U(4) representations for each species due to the electric charge superselection rule:

$$D_j = D_{je} + D_{jv} + D_{jupq} + D_{jdnq},$$

for j = 1, 2, 3, 4.

There are four layers of each species in the Normal and in the Dark matter sectors. The second, third and fourth layers of normal fermions has not as yet been found due to their extremely large masses.

A Layer Group rotates the fundamental fermions of each fundamental representation separately for each of the four species of each of the four generations.

The Layer Groups guarantee that all layers of each species have the same electric charge and other quantum numbers.

Each U(4) Layer Group also specifies a gauge field interaction among the fermions of its fundamental representation, species by species, for both Normal and Dark sectors. The form of the interactions is:

$$g_{ei}\overline{\Psi}_{ei}\gamma \cdot C_{ei} \cdot T\Psi_{ei} + g_{vi}\overline{\Psi}_{vi}\gamma \cdot C_{vi} \cdot T\Psi_{vi} + g_{upqi}\overline{\Psi}_{upqi}\gamma \cdot C_{upqi} \cdot T\Psi_{upqi} + g_{dnqi}\overline{\Psi}_{dnqi}\gamma \cdot C_{dnqi} \cdot T\Psi_{dnqi}$$

$$(11\text{-}A.2)$$

for i = generation = 1, … , 4, where g_{ei} … are coupling constants, the gauge fields are C_{ei} … , and the Ψ_{ei} … are 4-vectors of fermions formed of the ith generation fermions in each layer of each species.

The gauge vector bosons of the Layer Groups also have large masses. If the conservation of the fermion particle numbers is broken then we view it as a consequence of Layer Groups symmetry breaking.

Layer Group rotations guarantee the internal quantum numbers of each layer of each species are the same since symmetry breakdown is not present at the instant of the Big Bang.

The above discussion applies similarly to the Dark sector. There are 8 Layer Groups in NEWQUeST/NEWUST. See Figs. 11.1 and 11.2.

Fig. 11-A.1 shows the fundamental fermion spectrum with the representations of the Generation groups and Layer groups indicated.

Experimentally, we know of three generations of fermions—the lowest 3 generations of the lowest level. The remaining 4th generation and three layers of fermions are of much higher mass and are yet to be found.

See Blaha (2019g) and (2018e) for a detailed discussion of the Layer Groups. We note in passing that the symmetries of these number operators are badly broken. Yet the underlying group structure remains.

11-A.4 U(2) Connection Groups

The seven U(2) Connection groups of Fig. 11.3 generate rotations and interactions between *corresponding* fermions and vector bosons of each pair of blocks of the eight blocks of fermions in NEWQUeST/NEWUST.

Horizontal Lines

The horizontal lines in Fig. 11.3 (A, B, C, and D) each represent a U(2) Connection group that rotates two *corresponding* fermions in the Normal and Dark sectors of each layer. Thus a Normal e is rotated with a corresponding Dark e, and so on.

Each of the four horizontal Connection Groups has a reducible U(2) representation D that is the sum of 32 irreducible U(2) representations. We may view each reducible representation D as an array of 32 U(2) irreducible representations D_j strung along the diagonal.

$$D = \sum_{j=1}^{32} D_j$$

for each of the U(2) groups of the four horizontal lines in Fig. 11.3.

The U(2) group also specifies gauge field interactions between corresponding fermions in each layer of the Normal and Dark sectors of the form

$$g\overline{\Psi}_{Nn}\gamma\cdot A\cdot T\Psi_{Dn} \qquad (11\text{-}A.3)$$

where N indicates a Normal fermion and D indicates the corresponding Dark fermion, with A being a U(2) gauge vector boson, and n the label for corresponding fermions.

These U(2) transformations imply that the Normal and Dark sectors have the same species and the same set of internal symmetries fermion by fermion.

Vertical Lines

The pairs of vertical lines in Fig. 11.3 (E, F, G) each represent a U(2) Connection group that rotates sets of two *corresponding* fermions in adjacent layers as shown in Fig. 11.3 in the Normal and Dark sectors. Thus a Normal e in layer 1 is rotated with a corresponding Normal e in layer 2, and so on.

Each of the three (six counting both Normal and Dark lines in Fig. 11.3) vertical Connection Groups has a reducible U(2) representation D that is the sum of 64 irreducible U(2) representations. We may view each reducible representation D as an array of 64 U(2) irreducible representations D_j strung along the diagonal.

$$D = \sum_{j=1}^{64} D_j$$

for each of the U(2) groups of the 3 (6) horizontal lines in Fig. 11.3. Note the 64 irreducible representations include both Normal and Dark sectors of a layer. [41]

Each U(2) group also specifies a gauge field interaction between corresponding fermions in adjacent layers for both Normal and Dark sectors:

$$g\overline{\Psi}_{nl_1}\gamma\cdot A\cdot T\Psi_{nl_2} \tag{11-A.4}$$

where l_1 and l_2 designate layers, A is a gauge field vector boson, and n the label for corresponding fermions.

Each E, F, and G U(2) group reducible representation includes both Normal and Dark sectors.

11-A.5 A Unification of Symmetries in the NEWQUeST/NEWUST Fermion Spectrum

The Generation and Layer Groups are diagonal in the four fermion species. Fermions, considered species by species for all generations and layers in Normal and Dark sectors, have the same set of symmetries. Thus the Standard Model symmetries carry over directly to all parts of NEWQUeST/NEWUST.

If the Layer groups and the Connection groups were not present then each of the eight blocks might have differing sets of internal symmetries.

---

[41] There are 64 fermions in total for each of the four layers of NEWQUeST/NEWEST.

The Fermion Periodic Table

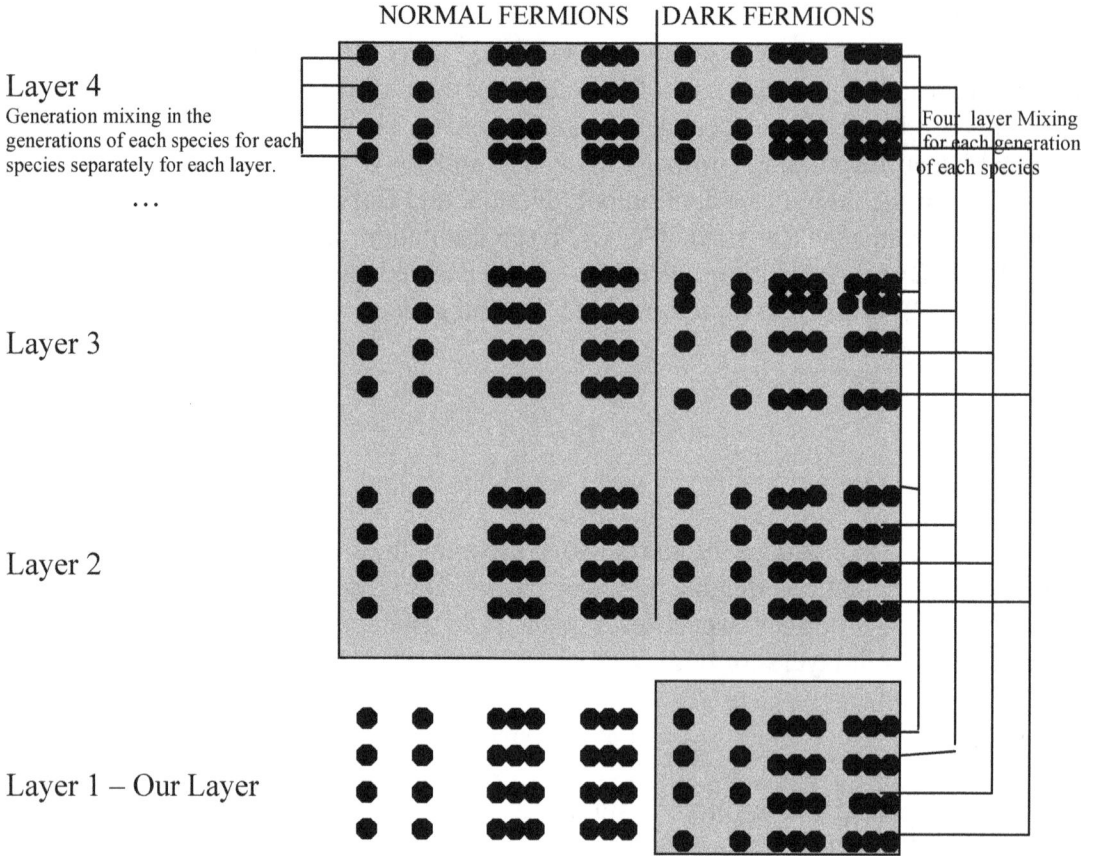

Figure 11-A.1. Fermion particle spectrum and partial examples of the pattern of mass mixing of the Generation groups and of the Layer groups. Unshaded parts are the known fermions with an additional, as yet not found, 4th generation. The lines on the left side (only shown for one layer) display the Generation mixing within each layer. The Generation mixing occurs within each layer using a separate Generation group for each layer. The lines on the right side show Layer group mixing (for Dark matter) with the mixing among all four layers for each of the four generations individually. There are four Layer groups for Normal matter and four Layer groups for Dark matter.. There are 256 fundamental fermions. NEWQUeST/NEWUST have the same fermion spectrum.

12.   NEWUTMOST (for the Megaverse)

In defining the UST, the author was motivated to add a Megaverse within which the universe resided. Based on the form of UST, and the form of QUeST, the UTMOST octonion space was defined. Subsequently, after finding the origin of QUeST in Cayley numbers and developing the Octonion Cosmology Spectrum described earlier and earlier books in 2020, UTMOST was found to be the Cayley number n = 5 space.[42]

Consequently, NEWUTMOST can be viewed as consisting of four copies of NEWQUeST. Thus due to the nature of Cayley numbers and the Octonion Spaces Spectrum, the total number of UTMOST/NEWUTMOST dimensions was $4 \cdot 256 = 1024$ dimension. The UTMOST/NEWUTMOST array was 32×32 in size and consists of four copies of the NEWQUeST array. The internal symmetries of UTMOST/NEWUTMOST are four copies of Figs. 11.1 and 11.2.

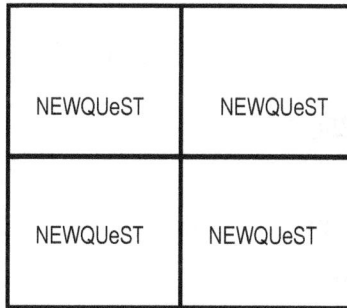

Figure 12.1. The NEWUTMOST dimension array viewed as composed of four copies of NEWQUeST.

The 16 space-time dimensions (four sets of four space-time dimensions from the four copies) can be combined to yield an UTMOST/NEWUTMOST U(4) Connection Group plus an eight real dimension space-time. Thus the total dimensions of the NEWUTMOST array is composed of four NEWQUeST sets of 252 internal symmetry dimensions[43] plus the eight dimensions of a Connection U(4) plus eight real space-time dimensions of NEWUTMOST giving a total of 1024 NEWUTMOST dimensions:

---

[42] **Upon the trsansfer of dimensions from space-time to a U(4) Connection Group (Fig. 12.6) we renamed UTMOST to NEWUTMOST. The NEWUTMOST array has four copies of the NEWQUeST array. Each NEWQUeST array has seven Connection Groups within it connecting its parts together.**
[43] NEWQUeST has four real space-time dimensions. These four NEWQUeST copies give 16 real space-time dimensions to NEWUTMOST. Eight of these 16 dimensions are transferred to form a U(4) Connection Group. The remainder becomes the 8 NEWUTMOST real space-time dimensions.

$$4 \cdot 252 \rightarrow 1008 + 8 \text{ U(4) dimensions} = 1016 \quad \text{NEWUTMOST internal symmetry dimensions}$$

$$+ 8 \quad \text{NEWUTMOST real space-time dimensions}$$

$$1024 \quad \text{Total NEWUTMOST dimensions}$$

The NEWUTMOST (and UTMOST) dimension array may be decomposed into blocks of 16 dimensions as shown in Fig. 12.2. Pairs of two blocks (32 dimensions) form Dimension-32 Atoms as described in earlier books. We go beyond the Normal-Dark sectors of NEWUST defining various additional "Dark" sectors: Normal+Dark1, Dark2+Dark3, Dark4+Dark5 and Dark6+Dark7.

Fig. 12.3 shows the breakdown of *one* NEWUTMOST layer into internal symmetry groups.

12.1 NEWUTMOST Internal Symmetries
The total internal symmetry is

$$\{[SU(2) \otimes U(1) \otimes SU(3) \otimes U(1)]^8 \otimes U(4)^{16} \otimes U(2)^7\}^4 \otimes U(4) \qquad (12.1)$$

Fig. 12.6 shows the four NEWQUeST parts of NEWUTMOST with the U(4) Connection Group. This group generates rotations among corresponding fermions (and gauge vector bosons) in the four parts. The group representation of the Connection U(4), D, is reducible. It has 256 irreducible fundamental U(4) representations D_i.

$$D = 256 D_i \qquad (12.2)$$

12.1.1 Space-Time Coordinates
Eight real space-time coordinates.

12.2 Fundamental Fermion Spectrum
There are $4 \cdot 256 = 1024$ fundamental fermions in NEWUTMOST. See Figs. 12.4 and 12.5. Fig. 12.5 shows the fermion spectrum in a Strong interaction SU(4) format. It could be changed to an SU(3)⊗U(1) format.

12.3 Total Dimensions
The total of internal symmetry and space-time dimensions is 1024.

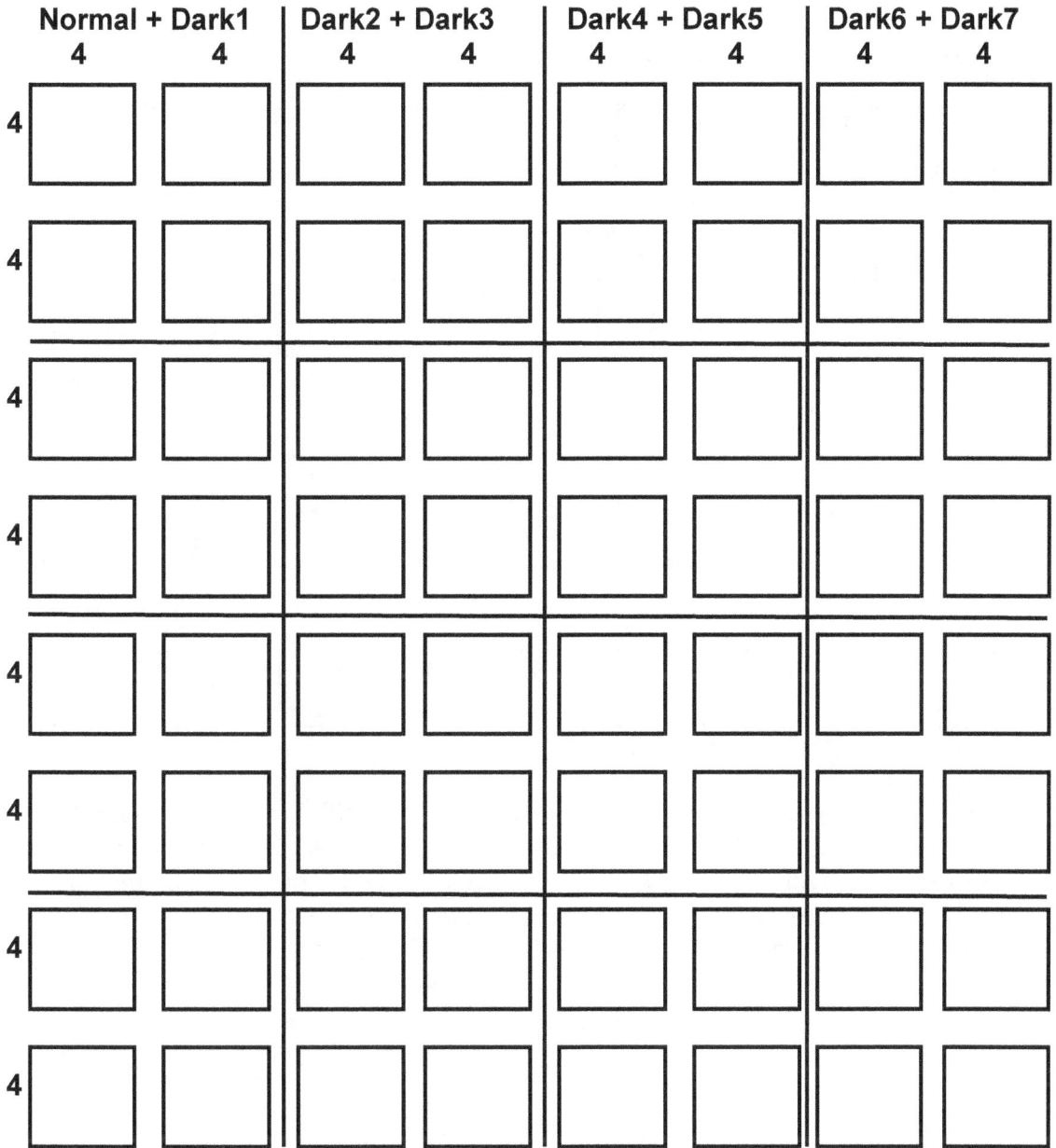

Figure 12.2. Four layers (each in two rows) in the 32 × 32 dimension NEWUTMOST array displayed as 4 × 4 = 16 dimension blocks counting space-time dimensions, some of which are later transferred to Connection groups. Each pair of these blocks form a Dimension-32 Atom. Two Dimension-32 Atoms form a 64 dimension blocks. Four Dimension-32 blocks form a 256 dimension layer.. The Dimension-32 Atoms pairs can be called sections: Normal+Dark1, Dark2+Dark3, Dark4+Dark5 and Dark6+Dark7 as shown above. In total they form the 32 × 32 = 1024 NEWUTMOST dimension array.

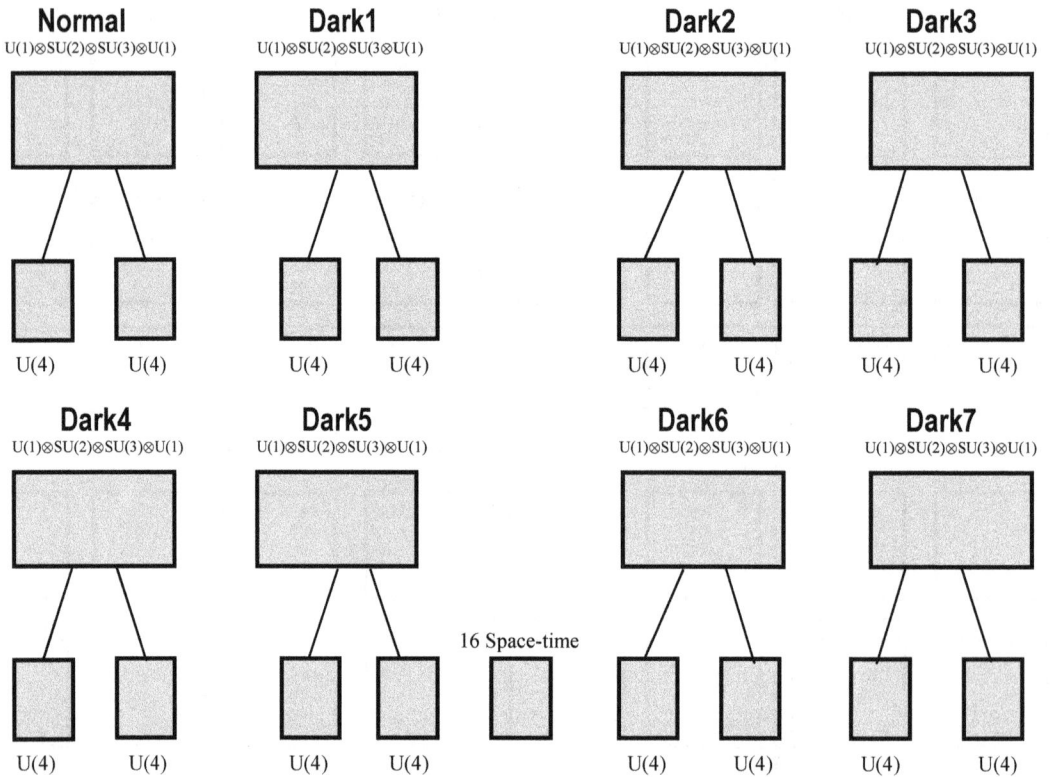

Figure 12.3. The internal symmetry groups of *one NEWUTMOST layer* of the four layers of 32 × 32 dimension NEWUTMOST. The other three layers are copies of this layer. Each U(1)⊗SU(2)⊗SU(3)⊗U(1) "block" displayed above has 12 dimensions because 4 of the 16 dimensions of a 16-block are space-time dimensions, which are transferred to form Connection Groups.

UTMOST Fermion Array

Normal	Dark1	Dark2	Dark3	Dark4	Dark5	Dark6	Dark7
••••••••	••••••••	••••••••	••••••••	••••••••	••••••••	••••••••	••••••••
••••••••	••••••••	••••••••	••••••••	••••••••	••••••••	••••••••	••••••••
••••••••	••••••••	••••••••	••••••••	••••••••	••••••••	••••••••	••••••••
••••••••	••••••••	••••••••	••••••••	••••••••	••••••••	••••••••	••••••••
••••••••	••••••••	••••••••	••••••••	••••••••	••••••••	••••••••	••••••••
••••••••	••••••••	••••••••	••••••••	••••••••	••••••••	••••••••	••••••••
••••••••	••••••••	••••••••	••••••••	••••••••	••••••••	••••••••	••••••••
••••••••	••••••••	••••••••	••••••••	••••••••	••••••••	••••••••	••••••••
••••••••	••••••••	••••••••	••••••••	••••••••	••••••••	••••••••	••••••••
••••••••	••••••••	••••••••	••••••••	••••••••	••••••••	••••••••	••••••••
••••••••	••••••••	••••••••	••••••••	••••••••	••••••••	••••••••	••••••••
••••••••	••••••••	••••••••	••••••••	••••••••	••••••••	••••••••	••••••••
••••••••	••••••••	••••••••	••••••••	••••••••	••••••••	••••••••	••••••••
••••••••	••••••••	••••••••	••••••••	••••••••	••••••••	••••••••	••••••••
••••••••	••••••••	••••••••	••••••••	••••••••	••••••••	••••••••	••••••••
••••••••	••••••••	••••••••	••••••••	••••••••	••••••••	••••••••	••••••••

Figure 12.4. Spectrum of UTMOST fermions in a 16×64 format. Each fermion is represented by a •..Each set of eight •.'s represents a charged lepton, a neutral lepton, three up-type quarks, and three down-type quarks. There are eight sets of four species in four generations which are in turn in 4 layers. There are 1024 fundamental fermions taking account of quark triplets. The 16×64 format could be changed to a 32×32 format without adverse physical consequences by "juggling" the Dark sectors. See Fig. 12.5.

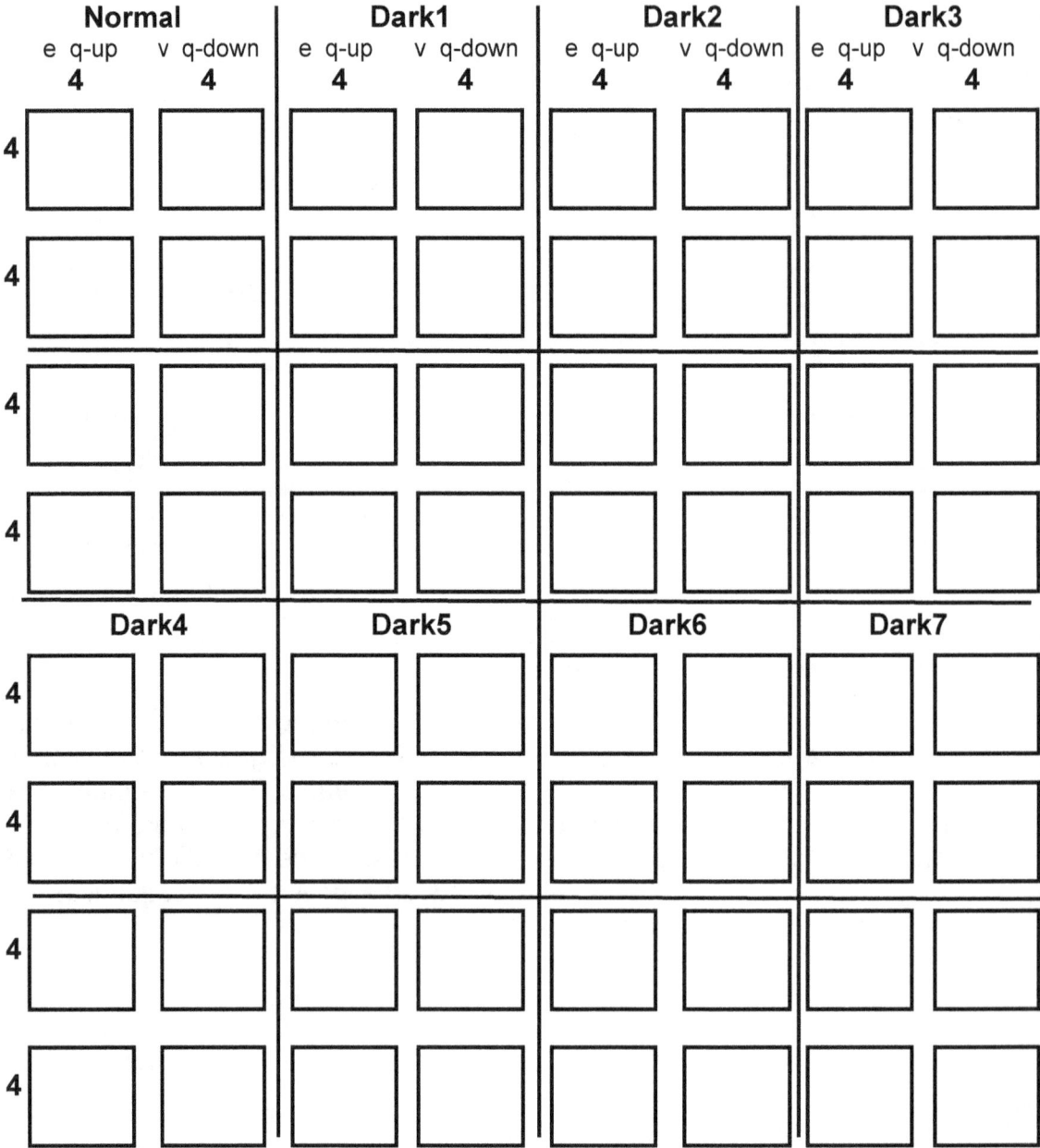

Figure 12.5. Block form of the 32 × 32 UTMOST fermion array with each row corresponding to *half of an UTMOST layer*. Thus 8 × ½ = 4 layers results. Each block contains four generations of fermions. The result is sixty-four 4 × 4 blocks. The label e q-up indicates a charged lepton – up-type quark pair, v q-down indicates a neutral lepton – down-type quark pair, and so on. *The form displayed here explains why generations come in fours.*

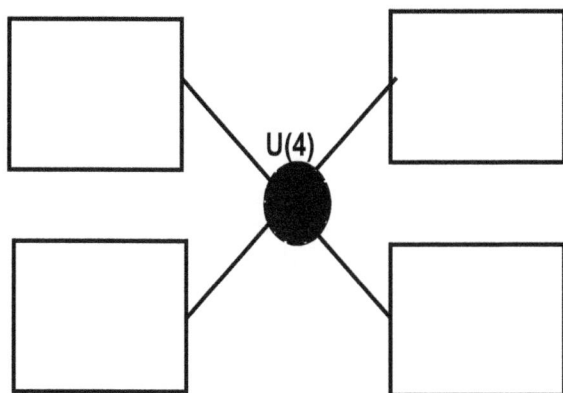

Figure 12.6. The U(4) Connection Group of NEWUTMOST.

13.   The Upper Limit of Octonion Spaces

The spectrum of Octonion Cosmology is specified in chapters 7 and 8 to have an end at the space with Cayley-Dickson number $n = 10$ having a 18 space-time dimension space within $1024 \times 1024 = 1,048,576$ total dimensions; From chapter 8 we have:

Octonion Space Number O_S	Cayley-Dickson Number n	Cayley Number Dimension d_c	Dimension Array Total d_d	Space-time-Dimension q	Fermion Spinor Array Total d_s	Cayley Number "Name"
0	10	1024	1024×1024	18	512×512	Complex Octonion Octonion Octonion

In eq. 7.1 we provisionally specified the $n = 10$ space as generated by a fermion-antifermion annihilation into a scalar particle Φ in a 20 space-time dimension parent space:

$$\Phi(20, 18, m_{20}, x) \tag{7.1}$$

We now consider the parent space with 20 space-time dimensions. A fermion space field has the form

$$\Psi_a(r, q, x) = \int d^{r-1}p(2\pi)^{-(r-1)}(m/p^0)^{\frac{1}{2}} \{\exp^{-ip \cdot x} U_{a\beta}(r, q, p)b_\beta(p) +$$

$$+ \exp^{ip \cdot x} V_{a\beta}(r, q, p)d^\dagger_\beta(p)\} \tag{1.30}$$

We will consider the possibility that the space-time is actually a point within 20 space-time dimensions. We will view the space-time as confined to a 20 dimension "box" that contracts to a point. As the box contracts a fermion space field integration becomes a discrete sum over k

$$\Psi_a(20, 18, x) = \Sigma_k (2\pi)^{-(r-1)}(m/p_k^0)^{\frac{1}{2}} \{\exp^{-ip_k \cdot x} U_{a\beta}(20, 18, p_k, m)b_\beta(p_k) +$$

$$+ \exp^{ip_k \cdot x} V_{a\beta}(20, 18, p_k, m)d^\dagger_\beta(p_k)\} \tag{13.1}$$

We now set

$$p_k = m \, p_k' \tag{13.2}$$
$$x = x'/m$$

in U and V, and take the limit to a point by letting $m \to \infty$. Due to simple dimensional scaling U and V are independent of m and dependent only on p_k'. They are non-zero in the point space limit.

$$\Psi_a(20, 18, x) = \Sigma_k (2\pi)^{-(r-1)}(p_k'^0)^{-\frac{1}{2}} \{\exp^{-ip_k'\cdot x'} U_{a\beta}(20, 18, p_k', 1)b_\beta(p_k') +$$

$$+ \exp^{ip_k'\cdot x'} V_{a\beta}(20, 18, p_k', 1)d^\dagger_\beta(p_k')\} \quad (13.3)$$

Thus in the point limit we obtain the 1024×1024 = 1,048,576 total dimension arrays from U and V. The n = 10 octonion space results with the dimension array partitioned into an 18 dimension space-time and 1,048,558 internal symmetry dimensions.

The lower spaces listed in Chapters 7 and 8 then follow from a sequence of fermion-antifermion annihilations. The collapse of the 20 dimension space-time precludes additional spaces above it.

We thus find a *point space* "source" for the Octonion Cosmology spectrum of spaces bounding the spectrum from above. The spectrum is bounded from below by the zero dimension space-time of the Cayley-Dickson n = 1 space.

Requiring the spaces to have dimension arrays with Cayley-Dickson number rows and columns constrains the space-time dimensions of the spectrum to be even integers ranging from 0 through 18. The Cayley-Dickson numbers of the spectrum range from n = 1 through n = 10. See Fig. 8.1.

We find the Octonion Cosmology spectrum of spaces arises from a point (of infinite energy m = ∞) in a 20 dimension space-time.

14. Experimental Implications of Octonion Cosmology

The clear connection between the spinor arrays generated by fermion-antifermion annihilation and the fermion spectrum and the internal symmetries (chapter 9 and Fig. 9.2 in particular) leads to decisive experimental tests of Octonion Cosmology.

Figs. 9.2a, 9.2b, and 9.2c below clearly show that spinor arrays structure the fundamental fermion spectrum and the set of internal symmetries of our universe as presented in NEWQUeST and NEWUST.

However, our current knowledge of the fermion spectrum and internal symmetries is very limited. See Figs. 14.1 and 14.2 below.

Therefore we suggest experimental tests of Octonion Cosmology including:

1. Finding a 4th generation of fundamental fermions.
2. Finding a second layer (and possibly more layers) of fermions and additional symmetries.
3. Finding the Dark Sector of fermions and symmetries.
4. Developing evidence for the Generation and Layer Groups through a reanalysis of current data as well as finding new data in the future.

If experiment yields some or all of the above, then there is strong support for Octonion Cosmology. At the moment Octonion Cosmology stands out as the only *complete* theory of Elementary Particle Physics and Cosmology.

Number of Columns =	4	4	4	4

	u-type fermion)		v-type (anti-fermion)	
4	u spin up		v small terms 1	
4		u spin down	v small terms 2	
4	u small terms 1		v spin down	
4		u small terms 2		v spin up

Figure 9.2a. The 16-spinors of a 8 space-time dimension (a Megaverse) spinor array. Each spinor column has 16 rows. There are 16 16-spinors.

SU(4) Fermion Particle Periodic Table

Number of Columns = 4   **NORMAL**   4   4   **DARK**   4

Layer 1 4	e 3 up-quarks u spin up	ν 3 down-quarks	e 3 up-quarks v small terms 1	ν 3 down-quarks
Layer 2 4	e 3 up-quarks	ν 3 down-quarks u spin down	e 3 up-quarks v small terms 2	ν 3 down-quarks
Layer 3 4	e 3 up-quarks u small terms 1	ν 3 down-quarks	e 3 up-quarks v spin down	ν 3 down-quarks
Layer 4 4	e 3 up-quarks u small terms 2	ν 3 down-quarks	e 3 up-quarks	ν 3 down-quarks v spin up

Figure 9.2b. The 16-spinors of a 8 space-time dimension spinor array. It becomes the dimension array of a universe. It corresponds directly with SU(4) **4** (or SU(3)⊗U(1)) fermions that are shown superimposed on the array. Each spinor column has 16 rows. There are 16 16-spinors. There are four layers. Each set of 4 fermions has 4 generations matching the number of rows in each layer. The Periodic Table is broken into Normal and Dark sectors. As shown above a Match!

WQUeST/NEWUST Symmetries Table

Number of Columns = 4   **NORMAL**   4   4   **DARK**   4

Layer 1 4	SU(4)⊗ Generation U(4) u spin up	SU(2)⊗U(1)⊗SL(2,C)[44] Layer U(4)	SU(4)⊗ Generation U(4) v small terms 1	SU(2)⊗U(1)⊗SL(2,C) Layer U(4)
Layer 2 4	SU(4)⊗ Generation U(4)	SU(2)⊗U(1)⊗SL(2,C) Layer U(4) u spin down	SU(4)⊗ Generation U(4) v small terms 2	SU(2)⊗U(1)⊗SL(2,C) Layer U(4)
Layer 3 4	SU(4)⊗ Generation U(4) u small terms 1	SU(2)⊗U(1)⊗SL(2,C) Layer U(4)	SU(4)⊗ Generation U(4) v spin down	SU(2)⊗U(1)⊗SL(2,C) Layer U(4)
Layer 4 4	SU(4)⊗ Generation U(4) u small terms 2	SU(2)⊗U(1)⊗SL(2,C) Layer U(4)	SU(4)⊗ Generation U(4)	SU(2)⊗U(1)⊗SL(2,C) Layer U(4) v spin up

Figure 9.2c. The16-spinors of a 8 space-time dimension spinor array. It becomes the dimension array of a universe. It corresponds directly with the pattern of symmetries of NEWUST. Note each layer has its own set of internal symmetries. Normal and Dark sector symmetries are also different. The Layer Groups and the Connection Groups tie the various parts together. As shown, the pattern of symmetries closely matches the NEWQUeST dimension array (spinor array) splittings.The 4 ×4 blocks match the symmetry splits SU(4) appears split into SU(3)⊗U(1) experimentally.The eight SL(2, **C**) groups are transformed into one 4-dimension space-time and seven U(2) connection groups. See Fig. 11.3 and section 11-A.4 for details

---

[44] The Lorentz group. $SO^+(1,3)$ in particular

The Fermion Periodic Table

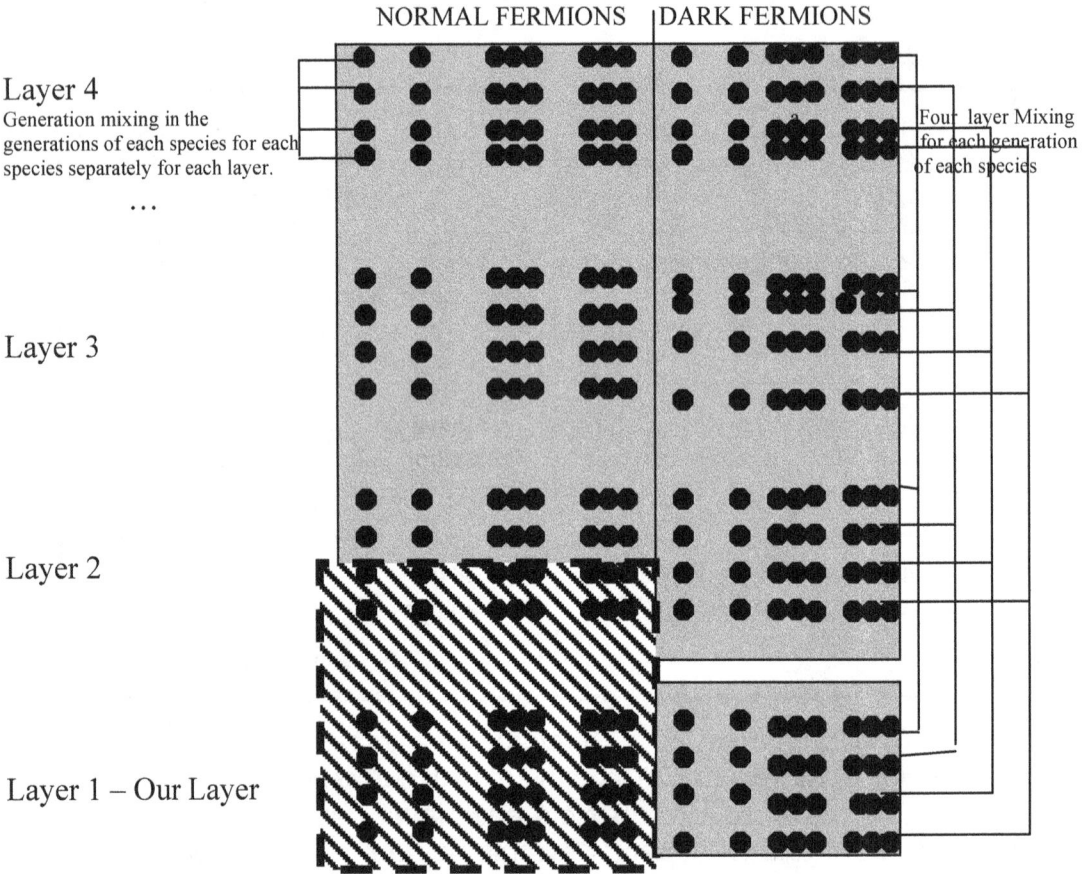

Figure 14.1. Fermion particle spectrum and partial examples of the pattern of mass mixing of the Generation groups and of the Layer groups. The cross hatched part are the known fermions and the desired *but not yet found* 4th generation and part of a second layer.

NEWQUeST Vector Bosons

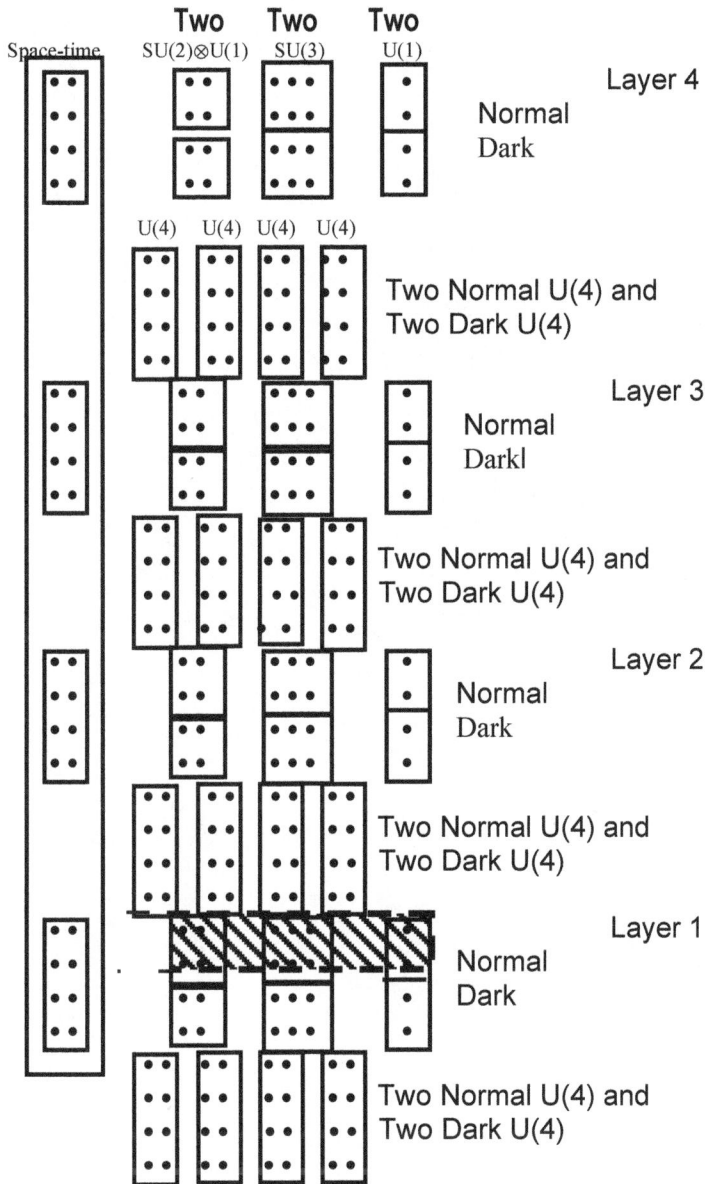

Figure 14.2. The four layers of QUeST internal symmetry groups (and space-time) for the 32 octonion dimension form of space. Note each layer requires 64 dimensions.. *Note the duplication (using the "Two" label) of each symmetry in each layer.The cross hatched part is the known interactions.*

15. Possible Relation to SuperString Theories, Composition Algebras

Seven of the ten spaces of Octonion Cosmology appear to correspond to SuperString Theories.

Cayley Number	Space-time Dimension	Spinor Size	Dimension Array Row Size, = Supercharge	Composition Algebra	SuperString Theory
4	6[45]	8	16	Complex Octonion	Type I
5	8	16	32	Quaternion Octonion	Type 2
6	10	32	64	Octonion Octonion	Heterotic?
7	12	64	128	Complex Octonion Octonion	?
8	14	128	256	Quaternion Octonion Octonion	?
9	16	256	512	Octonion Octonion Octonion	?
10	18	512	1024	Complex Octonion Octonion Octonion	?

The higher Cayley number spaces have yet to be matched to SuperString Theories. A possible match between Octonion Cosmology and SuperString theories may emerge.

The ? indicate additional theories suggested by the order of type I and type II supercharges, which also are the squares of Cayley numbers. Note the type I and type II SuperString Theories are assigned to different octonion spaces.

The comparison of Octonion Cosmology with its series of steps connecting to NEWQUeST and the Unified SuperStandard Theory (NEWUST) with the goal of SuperString Theory to connect to Elementary Particles suggests direct steps of SuperString Theory to Particle Theory are likely to be difficult. The possibility of SuperString Theory as a basis of Octonion Cosmology is evident.

---

[45] Possibly 4 with two compacted dimensions.

Appendix A. The 10 Spaces of Octonion Cosmology

This appendix contains chapter 1 (unchanged) of *From Octonion Cosmology to the Unified SuperStandard Theory of Particles* published December 27, 2020. It describes the 10 spaces of Octonion Cosmology based on a strict creation of spaces below the space of our universe via fermion-antifermion annihilation. Comparing it with the current view, which is expressed in the preceding chapters, we see that only modest changes in the details have been made. The current view is significantly more robust with its addition of new aspects. However, it does appear that Octonion Cosmology and NEWUST has achieved stability and maturity.

1. The Spectrum of 10 Octonion Spaces

In earlier books in 2020 (See References) we successfully based the Unified SuperStandard Theory (UST) on the Quaternion Unified SuperStandard Theory (QUeST) with a remarkable match between the internal symmetries and space-time symmetry of both theories.

We will now describe a Cosmology based on a spectrum of octonion spaces that includes interlocked universe and Megaverse instances (particles). This Cosmology appears to be at the deepest level of physical reality. In this chapter we outline the features of its ten spaces. We describe the details of the spaces and their interrelations in the chapters that follow.

1.1 Octonion Basis

The most general type of number that can support quantum field theory (and perturbation theory) are octonions. Octonions have significant properties that enable them to be used in a quantum field theory development:

0. An octonion is an 8-tuple of real numbers. A complex octonion is an 8-tuple of complex numbers.
2. Octonions are non-associative.
3. Octonions are one of the two finite dimensional division rings having the real numbers as a proper subring. (The other is quaternions—considered in chapter 5.)
4. Octonions are non-commutative. (This is not a roadblock for quantum field Theory, which is also non-commutative in general.)

We can represent a complex octonion (bioctonion) b as

$$b = b_{real} + Ib_{imaginary}$$

where b_{real} and $b_{imaginary}$ are real-valued octonions, and I is an additional fundamental octonion unit . We can also represent a complex octonion as an octonion with complex coordinates. Section 0.11 describes generalizations of octonion coordinates.

The octonion spaces in this book are based on treating octonions as the source of dimensions. Octonion units algebra is not used.

1.2 The Spectrum of Physical Octonion Spaces

There is a set of 10 conceptually interrelated spaces built on octonions. These spaces are listed in Fig. 1.1. Further types of octonion spaces can be defined. But they do not appear to be necessary, and their connection to the 10 spaces can be eliminated (truncated) by the specifications of the Complex Octonion Octonion[46] space (space 3) and the Quaternion Minispace space 10. We describe the limitation to 10 spaces in detail in later chapters.

OCTONION SPACES SPECTRUM

Spectrum Number

	Coordinate Type	Number of Coordinates	Dimension Array Size	Space-Time Dimensions
	Superverse			
0	Complex Octonion Octonion Octonion (1024)	Complex Octonion Octonion Octonion	1024 × 1024	0
	Spaceless[47]			
1	Octonion Octonion Octonion (512)	Octonion Octonion Octonion	512 × 512	0
2	Quaternion Octonion Octonion (256)	Quaternion Octonion Octonion	256 × 256	0
3	Complex Octonion Octonion (128)	Complex Octonion Octonion	128 × 128	0
	Cosmology[48]			
4	Octonion Octonion (64) Maxiverse	Octonion Octonion	64 × 64	10 quaternion octonion
5	Quaternion Octonion[49] (32) Megaverses	Quaternion Octonion	32 × 32	8 complex octonion
6	Complex Octonion[50] (16) Universes	Complex Octonion	16 × 16	4 octonion
	Minispaces[51]			
7	Quaternion (4)	Quaternion	4 × 4	4 Real
8	Real (4)	Real (4)	4 × 4	4 Real
9	Real (4)	Real (4)	4 × 4	4 Real
10	Real (4)	Real (4)	4 × 4	0

Figure 1.1. The spectrum of the Superverse and the ten octonion spaces. The spaces are numbered from 0 through 10. The numbers in parentheses in column 2 are the number of dimensions in each coordinate. The items in column 3 are the number of rows of dimensions (1024, 512, 256, 128, 64, 32, 16, 4, 4, 4,4).

1.3 Characteristics and Roles of the Spaces

The spaces of Octonion Cosmology have a variety of features that ultimately lead to the Unified SuperStandard Theory (UST). We describe the relationships in

---

[46] **This phrase, and similar phrases, are not typos.** The phrase "Octonion Octonion Octonion" means 8 octonion octonions, which means 64 octonions, which means a 512-tvector for a coordinate. This coordinate has 512 real values assembled in a 512-vector. Similar comments apply to other entries in Fig. 1.1. See chapter 0.

[47] Spaceless spaces are spaces without a space-time.

[48] Cosmological spaces are those, which are directly related to physics from megaverses and universes to elementary particles.

[49] In our earlier books in 2020 we also designated this 1024 dimension space (5) as 64 complex octonion space.

[50] In our earlier books in 2020 we also designated this 256 dimension space(6) as 32 complex quaternion space.

[51] A Minispace is a subspace of a universe space.

detail. UST bears direct comparison to experiment and accommodates the known features of The Standard Model of Elementary Particles.

The spectrum of spaces is also of interest in its own right.[52]

1.3.0 Superverse

The Superverse is a space with no space-time coordinates within it. It is formless since there is no dynamics and thus has no symmetry breaking. It contains spaces 1 through 10 as subspaces. See Fig. 1.3 and chapter 13.

1.3.1 Spaceless Spaces (1 – 3)

The three spaceless spaces of Fig. 1.1 are each defined to have dimensions that are not separated into subspaces. Thus there are no specified internal symmetry representations and no space-times within them.[53] There are no particles. There is no dynamics. There is no spatial separation within them—each space can be viewed as a point space. Only their "essence" exists. See chapter 11.

1.3.2 Octonion Octonion Maxiverse Space (4)

Instances of this space, the Maxiverse, cannot be constructed by the annihilation of spaceless space fermions since no fermions exist in the spaceless spaces. See Fig. 1.2. This space has 4096 dimensions, which includes a 10 quaternion octonion space-time. It has urfermions (a type of fermion described later) with 32-spinors. It has internal symmetries and bosons that result from the upward extension of Megaverse space internal symmetries (that extend, in term, from QUeST universe internal symmetries). It has dynamical evolution. It can create Megaverse space (5) instances through urfermion-antiurfermion annihilation. See chapter 3 for more details. Only one instance (particle) of Maxiverse space is created. See chapter 10.

1.3.3 Quaternion Octonion Megaverse Spaces (5)

Instances of this space can be constructed by the annihilation of octonion octonion space fermions. See Fig. 1.2. This space has 1024 dimensions, which includes an 8 complex octonion space-time. A Megaverse instance has a location and momentum in the octonion octonion (4)'s space.-time. It has fermions with 16-spinors. It has internal symmetries that result from the upward extension of QUeST universe internal symmetries. It experiences dynamical evolution. It can create QUeST universe spaces through fermion-antifermion annihilation. See chapter 6 for more details. Any number of instances (particles) of Megaverse space can be created. We know of one Megaverse at most.

---

[52] We explore these spaces from a physical perspective. It is also possible to view these spaces as analogues of Philosophical-Religious phenomena. The three spaceless spaces can be viewed as an analogue of a deity; and the 64 dimension Octonion Octonion space with 10 dimensions and 32-spinor fermions could be viewed as an analogue of the concept of 10 Sefirot with 32 emanations in Judaism. See the *Sefer Yetzirah* – The Book of Creation. Also see Appendix G.

[53] We use the absence of space-times to end (truncate) the sequence of spaces that are connected by fermion-antifermion annihilation as discussed later.

1.3.4 Complex Octonion Universe Spaces (6)

Instances of this space can be constructed by the annihilation of quaternion octonion (5) space fermions. See Fig. 1.2. This space has 256 dimensions, which includes a 4 octonions space-time. A universe instance has a location and momentum in a Megaverse's space.-time. It contains fermions with 4-spinors. It has QUeST internal symmetries (consistent with UST) It experiences dynamical evolution. It can create minispaces through fermion-antifermion annihilation. See chapter 2. Any number of instances (particles) of universe space can be created. We know of one universe at present.

1.3.5 Quaternion Point Universe Minispaces (7)

Instances of this space can be constructed by the annihilation of complex octonion space fermions in space 6. See Fig. 1.2. This point space has a real-valued 4-dimension space-time. See chapter 12. Any number of instances (particles) of this space can be created.

1.3.6 Quaternion Point Universe Minispaces (8 and 9)

Instances of space 8 can be constructed by the annihilation of complex octonion space fermions in space 7. See Fig. 1.2. This point space has a real-valued 4-dimension space-time. See chapter 12.

Space 9 is constructed by the annihilation of real space-time fermions in space 8.. See Fig. 1.2. This point space has a real-valued 4-dimension space-time. (chapter 12.) Any number of instances (particles) of this spaces can be created.

1.3.7 Quaternion Point Universe Minispaces (10)

Instances of this space can be constructed by the annihilation of fermions in space 9. See Fig. 1.2. This point space has no space-time, by construction, to preclude an infinite sequence of nested minispaces. A minispace has a location and momentum in a universe's space.-time. It has no fermions or bosons. It can have a subset of QUeST internal symmetries. It does not experience dynamical evolution. See chapter 12. Any number of instances (particles) of this space can be created.

1.3.8 Comments

Instances of spaces 7 – 10 would be created within a universe space instance. An instance of space 7 might be possible in a universe space (6) such as our universe, and might be detected experimentally. Instances of spaces 8 – 10 would appear to be unlikely to be produced or found

The dimension of spaces 7 – 10 total to 64. This number would be the dimension of an octonion coordinate space of octonion coordinates $8 \times 8 = 64$, thus completing the progressive regular pattern of spaces seen in spaces from 0 through 6. The array size 8×8 is not present in Fig. 1.1 because a type 6 universe has fermions with 4-spinors. A space of type 7 can only have instances generated by type 6 fermion-antifermion annihilation. Thus it has a 4×4 dimension array resulting.

If instances of space 7 were derived from space 6 by fermion-antifermion annihilation, then space 7 would have been

| 7 Octonion (8) | Octonion | 8×8 | An invalid alternative |

replacing the 7, 8, 9, and 10 spaces.

The regularity of the pattern of regularly increasing dimension array sizes by factors of 2, and of the progression of coordinate types expressed in terms of octonions, in Fig. 1.1 is very encouraging.

1.4 The Pattern of the Ten Spaces

Spaces 1 through 6 in Fig. 1.1 show an orderly progression of spaces based on the coordinate type and number of coordinates in the second and third columns. Instances of spaces 4 through 10 are linked through fermion-antifermion annihilation as we will see later. The spectrum of spaces is bound above and below by spaces with no space-times and thus no particles or dynamics or fermion-antifermion annihilation.

The three top spaces labeled 1, 2, and 3 are present to round out the coordinates up to "octonion octonion octonion" and to "fill" the block Superverse space of Fig. 1.3.

While spaces 7 and 8 are needed to limit the space specrtrum, spaces 9 and 10 are not required. However we found them useful to create a unified Superverse space consisting of all 10 spaces joined together.

1.4.1 Unification of the 10 Spaces in the Superverse

The 10 spaces are interconnected as shown in Figs. 1.1, 1.2 and 1.3. We will see the interconnections in detail later. If we consider the union of the 10 spaces we find the total number of dimensions totals to 349,504.

The set of 10 spaces can be viewed as subspaces of a complex octonion octonion octonion *Superverse* space with coordinates numbering complex octonion octonion octonion or 1024. Each of the 1024 dimension coordinates is a 1024-vector of the complex octonion octonion octonion type. Fig. 1.3 shows this Superverse space in block-diagonal format.

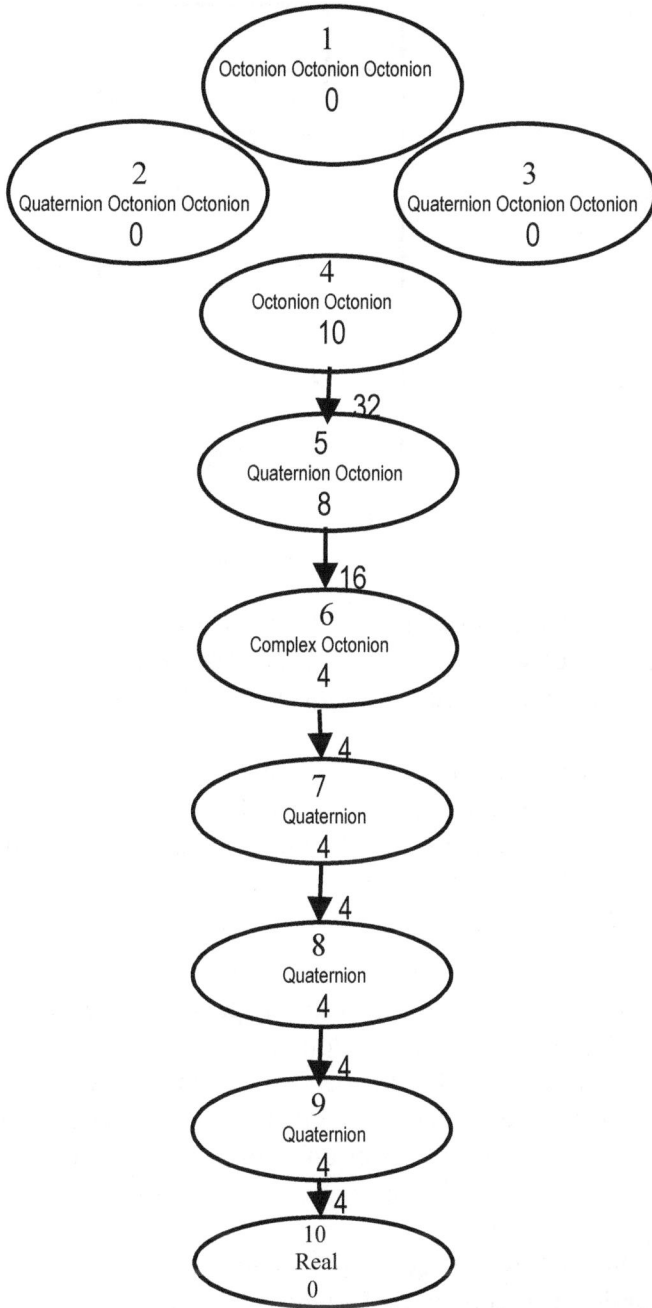

Figure 1.2. The descent of the ten spaces with their spectrum number and their space-time dimensions indicated within each oval. The number of spinor components for each fermion - antifermion pair that annihilates to produce the "next space down" is specified next to each arrow for the seven lower spaces.

Superverse

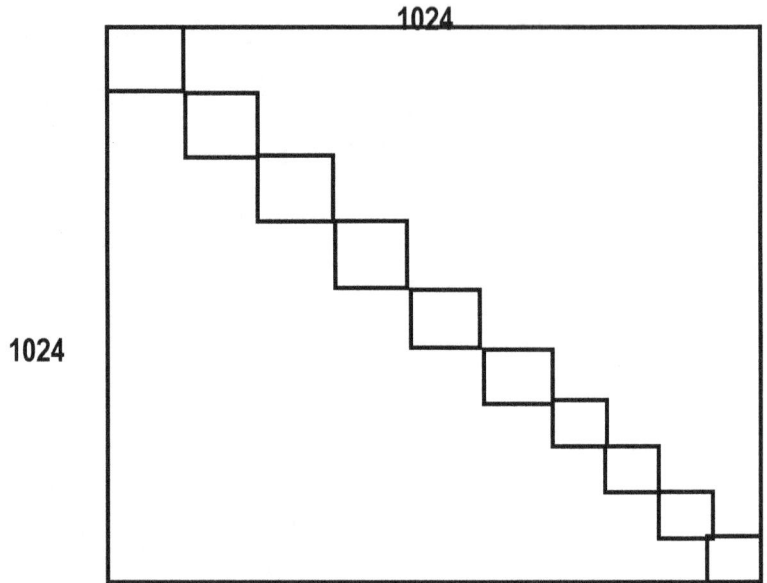

Figure 1.3. The 10 spaces of Fig. 1.1 in block form in a 1024 by 1024 dimension space – the Superverse space. The 10 spaces are 512 by 512, 256 by 256, ..., 4 by 4. The shapes of the spaces diagrammed in this figure is not significance.

The 1024×1024 Superverse is broken to the 10 spaces.

An analogy for the formation of the Cosmos based on the *Sefer Yetzirah* is presented in detail in Appendix G for the Superverse and its 10 subspaces. It provides a possible justification for Octonion Cosmology.

1.5 Cosmological Evolution

The spectrum of spaces has an evolutionary character. The 3 spaceless spaces are static. They serve to close the spectrum of spaces (an upper limit) while maintaining the mathematical trend of coordinate octonions.

The seven spaces 4 - 10 are needed for the physical Cosmos. The Complex Octonion Octonion (space 3) and the Minispace (space 10) mark the zero space-time dimension endpoints and thus the spectrum limits. These spaces are depicted in Fig. 1.4.

Turning now to the evolutionary sequence we find:

1. The Superverse space (type 0) exists without space-time, particles, or dynamics. It is broken to the tensor product of 10 subspaces.

2. Spaces 1, 2, and 3 are static without space-times, particles, or dynamics.

3. The Octonion Octonion (Maxiverse) Space instance must exist for all time. It is populated with elementary particles and interactions. After an unspecified time, an urfermion-antiurfermion annihilation may produce a

Megaverse space (type 5) instance. (There may be several/many Megaverse instances spread in time.)

4. A Megaverse space instance (particle) contains elementary particles and interactions. After an unspecified time a fermion-antifermion annihilation may produce a QUeST universe instance of type 6. There may be several/many universes over Megaverse time.

5. A QUeST universe contains the fermions, bosons and interactions of the Unified SuperStandard Theory.

6. Fermion-antifermion annihilation can produce a minispace (types 7, 8, 9) each containing a space-time, elementary particles and interactions..

7. Type 10 space has no space-time. Since it has no space-time it can have no particles or interactions. It is empty. It provides a lower limit to the spectrum of spaces. It also has topological significance.

Minispaces have not been found experimentally. Considering their presumed rarity, and their lack of an experimental signature, the lack of experimental confirmation is understandable. If minispaces were detected they would be significant evidence for Octonion Cosmology! Miniverses of types 7, 8, 9, and 10 could be produced by fermion-antifermion annihilation in principle but are extraordinarily unlikely to be produced.

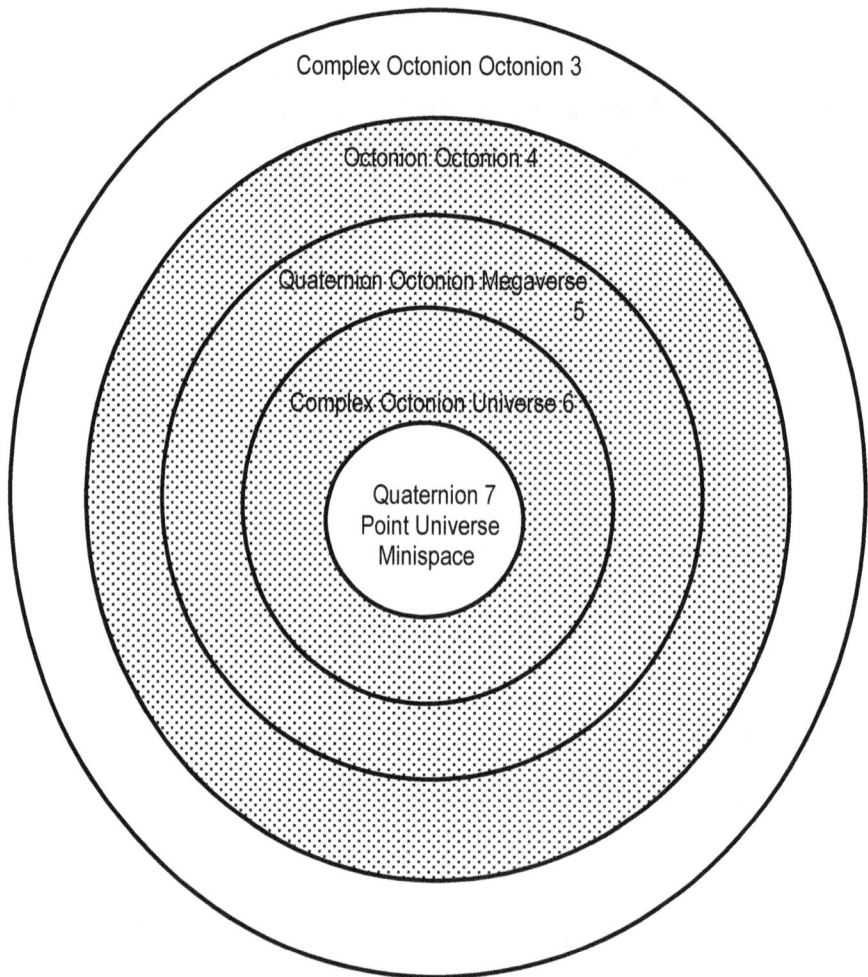

Figure 1.4. The five spaces relevant for the physical Cosmos. The three colored regions have non-zero dimension, and have a non-trivial space containing fermions and bosons with interactions.

REFERENCES

Akhiezer, N. I., Frink, A. H. (tr), 1962, *The Calculus of Variations* (Blaisdell Publishing, New York, 1962).

Bjorken, J. D., Drell, S. D., 1964, *Relativistic Quantum Mechanics* (McGraw-Hill, New York, 1965).

Bjorken, J. D., Drell, S. D., 1965, *Relativistic Quantum Fields* (McGraw-Hill, New York, 1965).

Blaha, S., 1995, *C++ for Professional Programming* (International Thomson Publishing, Boston, 1995).

_____, 1998, *Cosmos and Consciousness* (Pingree-Hill Publishing, Auburn, NH, 1998 and 2002).

_____, 2002, *A Finite Unified Quantum Field Theory of the Elementary Particle Standard Model and Quantum Gravity Based on New Quantum Dimensions™ & a New Paradigm in the Calculus of Variations* (Pingree-Hill Publishing, Auburn, NH, 2002).

_____, 2004, *Quantum Big Bang Cosmology: Complex Space-time General Relativity, Quantum Coordinates,™ Dodecahedral Universe, Inflation, and New Spin 0, ½, 1 & 2 Tachyons & Imagyons* (Pingree-Hill Publishing, Auburn, NH, 2004).

_____, 2005a, *Quantum Theory of the Third Kind: A New Type of Divergence-free Quantum Field Theory Supporting a Unified Standard Model of Elementary Particles and Quantum Gravity based on a New Method in the Calculus of Variations* (Pingree-Hill Publishing, Auburn, NH, 2005).

_____, 2005b, *The Metatheory of Physics Theories, and the Theory of Everything as a Quantum Computer Language* (Pingree-Hill Publishing, Auburn, NH, 2005).

_____, 2005c, *The Equivalence of Elementary Particle Theories and Computer Languages: Quantum Computers, Turing Machines, Standard Model, Superstring Theory, and a Proof that Gödel's Theorem Implies Nature Must Be Quantum* (Pingree-Hill Publishing, Auburn, NH, 2005).

_____, 2006a, *The Foundation of the Forces of Nature* (Pingree-Hill Publishing, Auburn, NH, 2006).

_____, 2006b, *A Derivation of ElectroWeak Theory based on an Extension of Special Relativity; Black Hole Tachyons; & Tachyons of Any Spin.* (Pingree-Hill Publishing, Auburn, NH, 2006).

_____, 2007a, *Physics Beyond the Light Barrier: The Source of Parity Violation, Tachyons, and A Derivation of Standard Model Features* (Pingree-Hill Publishing, Auburn, NH, 2007).

_____, 2007b, *The Origin of the Standard Model: The Genesis of Four Quark and Lepton Species, Parity Violation, the ElectroWeak Sector, Color SU(3), Three Visible Generations of Fermions, and One Generation of Dark Matter with Dark Energy* (Pingree-Hill Publishing, Auburn, NH, 2007).

_____, 2008a, *A Direct Derivation of the Form of the Standard Model From GL(16) (Pingree-Hill Publishing, Auburn, NH, 2008).*

_____, 2008b, *A Complete Derivation of the Form of the Standard Model With a New Method to Generate Particle Masses Second Edition* (Pingree-Hill Publishing, Auburn, NH, 2008)

_____, 2009, *The Algebra of Thought & Reality: The Mathematical Basis for Plato's Theory of Ideas, and Reality Extended to Include A Priori Observers and Space-Time Second Edition* (Pingree-Hill Publishing, Auburn, NH, 2009).

_____, 2010a, *Operator Metaphysics: A New Metaphysics Based on a New Operator Logic and a New Quantum Operator Logic that Lead to a Mathematical Basis for Plato's Theory of Ideas and Reality* (Pingree-Hill Publishing, Auburn, NH, 2010).

_____, 2010b, *The Standard Model's Form Derived from Operator Logic, Superluminal Transformations and GL(16)* (Pingree-Hill Publishing, Auburn, NH, 2010).

_____, 2010c, *SuperCivilizations: Civilizations as Superorganisms* (McMann-Fisher Publishing, Auburn, NH, 2010).

_____, 2011a, *21st Century Natural Philosophy Of Ultimate Physical Reality* (McMann-Fisher Publishing, Auburn, NH, 2011).

_____, 2011b, *All the Universe! Faster Than Light Tachyon Quark Starships & Particle Accelerators with the LHC as a Prototype Starship Drive Scientific Edition* (Pingree-Hill Publishing, Auburn, NH, 2011).

_____, 2011c, *From Asynchronous Logic to The Standard Model to Superflight to the Stars* (Blaha Research, Auburn, NH, 2011).

_____, 2012a, *From Asynchronous Logic to The Standard Model to Superflight to the Stars volume 2: Superluminal CP and CPT, U(4) Complex General Relativity and The Standard Model, Complex Vierbein General Relativity, Kinetic Theory, Thermodynamics* (Blaha Research, Auburn, NH, 2012).

_____, 2012b, *Standard Model Symmetries, And Four And Sixteen Dimension Complex Relativity; The Origin Of Higgs Mass Terms* (Blaha Reasearch, Auburn, NH, 2012).

_____, 2013a, *Multi-Stage Space Guns, Micro-Pulse Nuclear Rockets, and Faster-Than-Light Quark-Gluon Ion Drive Starships* (Blaha Research, Auburn, NH, 2013).

_____, 2013b, *The Bridge to Dark Matter; A New Sister Universe; Dark Energy; Inflatons; Quantum Big Bang; Superluminal Physics; An Extended Standard Model Based on Geometry* (Blaha Reasearch, Auburn, NH, 2013).

_____, 2014a, *Universes and Megaverses: From a New Standard Model to a Physical Megaverse; The Big Bang; Our Sister Universe's Wormhole; Origin of the Cosmological Constant, Spatial Asymmetry of the Universe, and its Web of Galaxies; A Baryonic Field between Universes and Particles; Megaverse Extended Wheeler-DeWitt Equation* (Blaha Reasearch, Auburn, NH, 2014).

_____, 2014b, *All the Megaverse! Starships Exploring the Endless Universes of the Cosmos Using the Baryonic Force* (Blaha Research, Auburn, NH, 2014).

_____, 2014c, *All the Megaverse! II Between Megaverse Universes: Quantum Entanglement Explained by the Megaverse Coherent Baryonic Radiation Devices – PHASERs Neutron Star Megaverse Slingshot*

Dynamics Spiritual and UFO Events, and the Megaverse Microscopic Entry into the Megaverse (Blaha Research, Auburn, NH, 2014).

_____, 2015a, *PHYSICS IS LOGIC PAINTED ON THE VOID: Origin of Bare Masses and The Standard Model in Logic, U(4) Origin of the Generations, Normal and Dark Baryonic Forces, Dark Matter, Dark Energy, The Big Bang, Complex General Relativity, A Megaverse of Universe Particles* (Blaha Research, Auburn, NH, 2015).

_____, 2015b, *PHYSICS IS LOGIC Part II: The Theory of Everything, The Megaverse Theory of Everything, U(4)⊗U(4) Grand Unified Theory (GUT), Inertial Mass = Gravitational Mass, Unified Extended Standard Model and a New Complex General Relativity with Higgs Particles, Generation Group Higgs Particles* (Blaha Research, Auburn, NH, 2015).

_____, 2015c, *The Origin of Higgs ("God") Particles and the Higgs Mechanism: Physics is Logic III, Beyond Higgs – A Revamped Theory With a Local Arrow of Time, The Theory of Everything Enhanced, Why Inertial Frames are Special, Universes of the Mind* (Blaha Research, Auburn, NH, 2015).

_____, 2015d, *The Origin of the Eight Coupling Constants of The Theory of Everything: U(8) Grand Unified Theory of Everything (GUTE), S^8 Coupling Constant Symmetry, Space-Time Dependent Coupling Constants, Big Bang Vacuum Coupling Constants, Physics is Logic IV* (Blaha Research, Auburn, NH, 2015).

_____, 2016a, *New Types of Dark Matter, Big Bang Equipartition, and A New U(4) Symmetry in the Theory of Everything: Equipartition Principle for Fermions, Matter is 83.33% Dark, Penetrating the Veil of the Big Bang, Explicit QFT Quark Confinement and Charmonium, Physics is Logic V* (Blaha Research, Auburn, NH, 2016).

_____, 2016b, *The Periodic Table of the 192 Quarks and Leptons in The Theory of Everything: The U(4) Layer Group, Physics is Logic VI* (Blaha Research, Auburn, NH, 2016).

_____, 2016c, *New Boson Quantum Field Theory, Dark Matter Dynamics, Dark Matter Fermion Layer Mixing, Genesis of Higgs Particles, New Layer Higgs Masses, Higgs Coupling Constants, Non-Abelian Higgs Gauge Fields, Physics is Logic VII* (Blaha Research, Auburn, NH, 2016).

_____, 2016d, *Unification of the Strong Interactions and Gravitation: Quark Confinement Linked to Modified Short-Distance Gravity; Physics is Logic VIII* (Blaha Research, Auburn, NH, 2016).

_____, 2016e, *MoND: Unification of the Strong Interactions and Gravitation II, Quark Confinement Linked to Large-Scale Gravity, Physics is Logic IX* (Blaha Research, Auburn, NH, 2016).

_____, 2016f, *CQ Mechanics: A Unification of Quantum & Classical Mechanics, Quantum/Semi-Classical Entanglement, Quantum/Classical Path Integrals, Quantum/Classical Chaos* (Blaha Research, Auburn, NH, 2016).

_____, 2016g, *GEMS: Unified Gravity, ElectroMagnetic and Strong Interactions: Manifest Quark Confinement, A Solution for the Proton Spin Puzzle, Modified Gravity on the Galactic Scale* (Pingree Hill Publishing, Auburn, NH, 2016).

_____, 2016h, *Unification of the Seven Boson Interactions based on the Riemann-Christoffel Curvature Tensor* (Pingree Hill Publishing, Auburn, NH, 2016).

_____, 2017a, *Unification of the Eleven Boson Interactions based on 'Rotations of Interactions'* (Pingree Hill Publishing, Auburn, NH, 2017).

_____, 2017b, *The Origin of Fermions and Bosons, and Their Unification* (Pingree Hill Publishing, Auburn, NH, 2017).

_____, 2017c, *Megaverse: The Universe of Universes* (Pingree Hill Publishing, Auburn, NH, 2017).

_____, 2017d, *SuperSymmetry and the Unified SuperStandard Model* (Pingree Hill Publishing, Auburn, NH, 2017).

_____, 2017e, *From Qubits to the Unified SuperStandard Model with Embedded SuperStrings: A Derivation* (Pingree Hill Publishing, Auburn, NH, 2017).

_____, 2017f, *The Unified SuperStandard Model in Our Universe and the Megaverse: Quarks, ... ,* (Pingree Hill Publishing, Auburn, NH, 2017).

_____, 2018a, *The Unified SuperStandard Model and the Megaverse SECOND EDITION A Deeper Theory based on a New Particle Functional Space that Explicates Quantum Entanglement Spookiness (Volume 1)* (Pingree Hill Publishing, Auburn, NH, 2018).

_____, 2018b, *Cosmos Creation: The Unified SuperStandard Model, Volume 2, SECOND EDITION* (Pingree Hill Publishing, Auburn, NH, 2018).

_____, 2018c, *God Theory (*Pingree Hill Publishing, Auburn, NH, 2018).

_____, 2018d, *Immortal Eye: God Theory: Second Edition* (Pingree Hill Publishing, Auburn, NH, 2018).

_____, 2018e, *Unification of God Theory and Unified SuperStandard Model THIRD EDITION* (Pingree Hill Publishing, Auburn, NH, 2018).

_____, 2019a, *Calculation of: QED α = 1/137, and Other Coupling Constants of the Unified SuperStandard Theory* (Pingree Hill Publishing, Auburn, NH, 2019).

_____, 2019b, *Coupling Constants of the Unified SuperStandard Theory SECOND EDITION* (Pingree Hill Publishing, Auburn, NH, 2019).

_____, 2019c, *New Hybrid Quantum Big_Bang–Megaverse_Driven Universe with a Finite Big Bang and an Increasing Hubble Constant* (Pingree Hill Publishing, Auburn, NH, 2019).

_____, 2019d, *The Universe, The Electron and The Vacuum* (Pingree Hill Publishing, Auburn, NH, 2019).

_____, 2019e, *Quantum Big Bang – Quantum Vacuum Universes (Particles)* (Pingree Hill Publishing, Auburn, NH, 2019).

_____, 2019f, *The Exact QED Calculation of the Fine Structure Constant Implies ALL 4D Universes have the Same Physics/Life Prospects* (Pingree Hill Publishing, Auburn, NH, 2019).

_____, 2019g, *Unified SuperStandard Theory and the SuperUniverse Model: The Foundation of Science* (Pingree Hill Publishing, Auburn, NH, 2019).

_____, 2020a, *Quaternion Unified SuperStandard Theory (The QUeST) and Megaverse Octonion SuperStandard Theory (MOST)* (Pingree Hill Publishing, Auburn, NH, 2020).

_____, 2020b, *United Universes Quaternion Universe - Octonion Megaverse* (Pingree Hill Publishing, Auburn, NH, 2020).

_____, 2020c, *Unified SuperStandard Theories for Quaternion Universes & The Octonion Megaverse* (Pingree Hill Publishing, Auburn, NH, 2020).

_____, 2020d, *The Essence of Eternity: Quaternion & Octonion SuperStandard Theories* (Pingree Hill Publishing, Auburn, NH, 2020).

_____, 2020e, *The Essence of Eternity II* (Pingree Hill Publishing, Auburn, NH, 2020).

_____, 2020f, *A Very Conscious Universe* (Pingree Hill Publishing, Auburn, NH, 2020).

_____, 2020g, *Hypercomplex Universe* (Pingree Hill Publishing, Auburn, NH, 2020).

_____, 2020h, *Beneath the Quaternion Universe* (Pingree Hill Publishing, Auburn, NH, 2020).

_____, 2020i, *Why is the Universe Real? From Quaternion & Octonion to Real Coordinates* (Pingree Hill Publishing, Auburn, NH, 2020).

_____, 2020j, *The Origin of Universes: of Quaternion Unified SuperStandard Theory (QUeST); and of the Octonion Megaverse (UTMOST)* (Pingree Hill Publishing, Auburn, NH, 2020).

_____, 2020k, *The Seven Spaces of Creation: Octonion Cosmology* (Pingree Hill Publishing, Auburn, NH, 2020).

_____, 2020l, *From Octonion Cosmology to the Unified SuperStandard Theory of Particles* (Pingree Hill Publishing, Auburn, NH, 2020).

_____, 2021a, *Pioneering the Cosmos* (Pingree Hill Publishing, Auburn, NH, 2021).

_____, 2021b, *Pioneering the Cosmos II* (Pingree Hill Publishing, Auburn, NH, 2021).

_____, 2021c, *Beyond Octonion Cosmology* (Pingree Hill Publishing, Auburn, NH, 2021).

_____, 2021d, *Universes are Particles* (Pingree Hill Publishing, Auburn, NH, 2021).

_____, 2021e, *Octonion-like dna-based life, Universe expansion is decay, Emerging New Physics* (Pingree Hill Publishing, Auburn, NH, 2021).

_____, 2021f, *The Science of Creation New Quantum Field Theory of Spaces* (Pingree Hill Publishing, Auburn, NH, 2021).

Eddington, A. S., 1952, *The Mathematical Theory of Relativity* (Cambridge University Press, Cambridge, U.K., 1952).

Fant, Karl M., 2005, *Logically Determined Design: Clockless System Design With NULL Convention Logic* (John Wiley and Sons, Hoboken, NJ, 2005).

Feinberg, G. and Shapiro, R., 1980, *Life Beyond Earth: The Intelligent Earthlings Guide to Life in the Universe* (William Morrow and Company, New York, 1980).

Gelfand, I. M., Fomin, S. V., Silverman, R. A. (tr), 2000, *Calculus of Variations* (Dover Publications, Mineola, NY, 2000).

Giaquinta, M., Modica, G., Souchek, J., 1998, *Cartesian Coordinates in the Calculus of Variations* Volumes I and II (Springer-Verlag, New York, 1998).

Giaquinta, M., Hildebrandt, S., 1996, *Calculus of Variations* Volumes I and II (Springer-Verlag, New York, 1996).

Gradshteyn, I. S. and Ryzhik, I. M., 1965, *Table of Integrals, Series, and Products* (Academic Press, New York, 1965).

Heitler, W., 1954, *The Quantum Theory of Radiation* (Claendon Press, Oxford, UK, 1954).

Huang, Kerson, 1992, *Quarks, Leptons & Gauge Fields 2nd Edition* (World Scientific Publishing Company, Singapore, 1992).

Jost, J., Li-Jost, X., 1998, *Calculus of Variations* (Cambridge University Press, New York, 1998).

Kaku, Michio, 1993, *Quantum Field Theory*, (Oxford University Press, New York, 1993).

Kirk, G. S. and Raven, J. E., 1962, *The Presocratic Philosophers* (Cambridge University Press, New York, 1962).

Landau, L. D. and Lifshitz, E. M., 1987, *Fluid Mechanics 2nd Edition*, (Pergamon Press, Elmsford, NY, 1987).

Misner, C. W., Thorne, K. S., and Wheeler, J. A., 1973, *Gravitation* (W. H. Freeman, New York, 1973).

Rescher, N., 1967, *The Philosophy of Leibniz* (Prentice-Hall, Englewood Cliffs, NJ, 1967).

Rieffel, Eleanor and Polak, Wolfgang, 2014, *Quantum Computing* (MIT Press, Cambridge, MA, 2014).

Riesz, Frigyes and Sz.-Nagy, Béla, 1990, *Functional Analysis* (Dover Publications, New York, 1990).

Sagan, H., 1993, *Introduction to the Calculus of Variations* (Dover Publications, Mineola, NY, 1993).

Sakurai, J. J., 1964, *Invariance Principles and Elementary Particles* (Princeton University Press, Princeton, NJ, 1964).

Weinberg, S., 1972, *Gravitation and Cosmology* (John Wiley and Sons, New York, 1972).

Weinberg, S., 1995, *The Quantum Theory of Fields Volume I* (Cambridge University Press, New York, 1995).

REFERENCES

INDEX

104

About the Author

Stephen Blaha is a well-known Physicist and Man of Letters with interests in Science, Society and civilization, the Arts, and Technology. He had an Alfred P. Sloan Foundation scholarship in college. He received his Ph.D. in Physics from Rockefeller University. He has served on the faculties of several major universities. He was also a Member of the Technical Staff at Bell Laboratories, a manager at the Boston Globe Newspaper, a Director at Wang Laboratories, and President of Blaha Software Inc. and of Janus Associates Inc. (NH).

Among other achievements he was a co-discoverer of the "r potential" for heavy quark binding developing the first (and still the only demonstrable) non-Aeolian gauge theory with an "r" potential; first suggested the existence of topological structures in superfluid He-3; first proposed Yang-Mills theories would appear in condensed matter phenomena with non-scalar order parameters; first developed a grammar-based formalism for quantum computers and applied it to elementary particle theories; first developed a new form of quantum field theory without divergences (thus solving a major 60 year old problem that enabled a unified theory of the Standard Model and Quantum Gravity without divergences to be developed); first developed a formulation of complex General Relativity based on analytic continuation from real space-time; first developed a generalized non-homogeneous Robertson-Walker metric that enabled a quantum theory of the Big Bang to be developed without singularities at t = 0; first generalized Cauchy's theorem and Gauss' theorem to complex, curved multi-dimensional spaces; received Honorable Mention in the Gravity Research Foundation Essay Competition in 1978; first developed a physically acceptable theory of faster-than-light particles; first derived a composition of extremums method in the Calculus of Variations; first quantitatively suggested that inflationary periods in the history of the universe were not needed; first proved Gödel's Theorem implies Nature must be quantum; provided a new alternative to the Higgs Mechanism, and Higgs particles, to generate masses; first showed how to resolve logical paradoxes including Gödel's Undecidability Theorem by developing Operator Logic and Quantum Operator Logic; first developed a quantitative harmonic oscillator-like model of the life cycle, and interactions, of civilizations; first showed how equations describing superorganisms also apply to civilizations. A recent book shows his theory applies successfully to the past 14 years of history and to *new* archaeological data on Andean and Mayan civilizations as well as Early Anatolian and Egyptian civilizations.

He first developed an axiomatic derivation of the form of The Standard Model from geometry – space-time properties – The Unified SuperStandard Model. It unifies all the known forces of Nature. It also has a Dark Matter sector that includes a Dark ElectroWeak sector with Dark doublets and Dark gauge interactions. It uses quantum coordinates to remove infinities that crop up in most

interacting quantum field theories and additionally to remove the infinities that appear in the Big Bang and generate inflationary growth of the universe. It shows gravity has a MOND-like form without sacrificing Newton's Laws. It relates the interactions of the MOND-like sector of gravity with the r-potential of Quark Confinement. The axioms of the theory lead to the question of their origin. We suggest in the preceding edition of this book it can be attributed to an entity with God-like properties. We explore these properties in "God Theory" and show they predict that the Cosmos exists forever although individual universes (or incarnations of our universe) "come and go." Several other important results emerge from God Theory such a functionally triune God. The Unified SuperStandard Theory has many other important parts described in the Current Edition of *The Unified SuperStandard Theory* and expanded in subsequent volumes.

Blaha has had a major impact on a succession of elementary particle theories: his Ph.D. thesis (1970), and papers, showed that quantum field theory calculations to all orders in ladder approximations could not give scaling deep inelastic electron-nucleon scattering. He later showed the eigenvalue equation for the fine structure constant α in Johnson-Baker-Willey QED had a zero at $\alpha = 1$ not 1/137 by solving the Schwinger-Dyson equations to all orders in an approximation that agreed with exact results to 4th order in α thus ending interest in this theory. In 1979 at Prof. Ken Johnson's (MIT) suggestion he calculated the proton-neutron mass difference in the MIT bag model and found the result had the wrong sign reducing interest in the bag model. These results all appear in Physical Review papers. In the 2000's he repeatedly pointed out the shortcomings of SuperString theory and showed that The Standard Model's form could be derived from space-time geometry by an extension of Lorentz transformations to faster than light transformations. This deeper space-time basis greatly increases the possibility that it is part of THE fundamental theory. Recently, Blaha showed that the Weak interactions differed significantly from the Strong, electromagnetic and gravitation interactions in important respects while these interactions had similar features, and suggested that ElectroWeak theory, which is essentially a glued union of the Weak interactions and Electromagnetism, possibly modulo unknown Higgs particle features, be replaced by a unified theory of the other interactions combined with a stand-alone Weak interaction theory. Blaha also showed that, if Charmonium calculations are taken seriously, the Strong interaction coupling constant is only a factor of five larger than the electromagnetic coupling constant, and thus Strong interaction perturbation theory would make sense and yield physically meaningful results.

In graduate school (1965-71) he wrote substantial papers in elementary particles and group theory: The Inelastic E- P Structure Functions in a Gluon Model. Phys. Lett. B40:501-502,1972; Deep-Inelastic E-P Structure Functions In A Ladder Model With Spin 1/2 Nucleons, Phys.Rev. D3:510-523,1971; Continuum Contributions To The Pion Radius, Phys. Rev. 178:2167-2169,1969; Character Analysis of U(N) and SU(N), J. Math. Phys. 10, 2156 (1969); and The Calculation of the Irreducible Characters of the Symmetric Group in Terms of the

Compound Characters, (Published as Blaha's Lemma in D. E. Knuth's book: *The Art of Computer Programming Vols. 1 – 4*).

In the early 1980's Blaha was also a pioneer in the development of UNIX for financial, scientific and Internet applications: benchmarked UNIX versions showing that block size was critical for UNIX performance, developing financial modeling software, starting database benchmarking comparison studies, developing Internet-like UNIX networking (1982) and developing a hybrid shell programming technique (1982) that was a precursor to the PERL programming language. He was also the manager of the AT&T ten-year future products development database. His work helped lead to commercial UNIX on computers such as Sun Micros, IBM AIX minis, and Apple computers.

In the 1980's he pioneered the development of PC Desktop Publishing on laser printers and was nominated for three "Awards for Technical Excellence" in 1987 by PC Magazine for PC software products that he designed and developed.

Recently he has developed a theory of Megaverses – actual universes of which our universe is one – with quantum particle-like properties based on the Wheeler-DeWitt equation of Quantum Gravity. He has developed a theory of a baryonic force, which had been conjectured many years ago, and estimated the strength of the force based on discrepancies in measurements of the gravitational constant G. This force, operative in D-dimensional space, can be used to escape from our universe in "uniships" which are the equivalent of the faster-than-light starships proposed in the author's earlier books. Thus travel to other universes, as well as to other stars is possible.

Blaha also considered the complexified Wheeler-DeWitt equation and showed that its limitation to real-valued coordinates and metrics generated a Cosmological Constant in the Einstein equations.

The author has also recently written a series of books on the serious problems of the United States and their solution as well as a book on the decline of Mankind that will follow from current social and genetic trends in Mankind.

In the past twenty years Dr. Blaha has written over 80 books on a wide range of topics. Some recent major works are: *From Asynchronous Logic to The Standard Model to Superflight to the Stars, All the Universe!, SuperCivilizations: Civilizations as Superorganisms, America's Future: an Islamic Surge, ISIS, al Qaeda, World Epidemics, Ukraine, Russia-China Pact, US Leadership Crisis, The Rises and Falls of Man – Destiny – 3000 AD: New Support for a Superorganism MACRO-THEORY of CIVILIZATIONS From CURRENT WORLD TRENDS and NEW Peruvian, Pre-Mayan, Mayan, Anatolian, and Early Egyptian Data, with a Projection to 3000 AD,* and *Mankind in Decline: Genetic Disasters, Human-Animal Hybrids, Overpopulation, Pollution, Global Warming, Food and Water Shortages, Desertification, Poverty, Rising Violence, Genocide, Epidemics, Wars, Leadership Failure.*

He has taught approximately 4,000 students in undergraduate, graduate, and postgraduate corporate education courses primarily in major universities, and large companies and government agencies.

Recently he developed a quantum theory, The Unified SuperStandard Theory (UST), which describes elementary particles in detail without the difficulties of conventional quantum field theory. He found that the internal symmetries of this theory could be exactly derived from an octonion theory called QUeST. He further found that another octonion theory (UTMOST) describes the Megaverse. It can hold QUeST universes such as our own universe. It has an internal symmetry structure which is a superset of the QUeST internal symmetries. This book provides a complete derivation from the Quantum Field Theory of Spaces.

www.ingramcontent.com/pod-product-compliance
Lightning Source LLC
Chambersburg PA
CBHW082008190326
41458CB00010B/3120

* 9 7 8 1 7 3 7 2 6 4 0 4 0 *